Principles of Data Wrangling
Practical Techniques for Data Preparation

Tye Rattenbury, Joseph M. Hellerstein, Jeffrey Heer,
Sean Kandel, and Connor Carreras

Beijing · Boston · Farnham · Sebastopol · Tokyo

Principles of Data Wrangling

by Tye Rattenbury, Joseph M. Hellerstein, Jeffrey Heer, Sean Kandel, and Connor Carreras

Copyright © 2017 Trifacta, Inc. All rights reserved.

Printed in the United States of America.

Published by O'Reilly Media, Inc., 1005 Gravenstein Highway North, Sebastopol, CA 95472.

O'Reilly books may be purchased for educational, business, or sales promotional use. Online editions are also available for most titles (*http://oreilly.com/safari*). For more information, contact our corporate/institutional sales department: 800-998-9938 or *corporate@oreilly.com*.

Editor: Shannon Cutt
Production Editor: Kristen Brown
Copyeditor: Bob Russell, Octal Publishing, Inc.
Proofreader: Christina Edwards

Interior Designer: David Futato
Cover Designer: Karen Montgomery
Illustrator: Rebecca Demarest

May 2017:   First Edition

Revision History for the First Edition
2017-04-25:   First Release
2017-06-27:   Second Release

The O'Reilly logo is a registered trademark of O'Reilly Media, Inc. *Principles of Data Wrangling*, the cover image, and related trade dress are trademarks of O'Reilly Media, Inc.

978-1-491-93892-8

[LSI]

Table of Contents

Foreword. vii

1. Introduction. 1
   Magic Thresholds, PYMK, and User Growth at Facebook   3

2. A Data Workflow Framework. 7
   How Data Flows During and Across Projects   8
   Connecting Analytic Actions to Data Movement: A Holistic Workflow
      Framework for Data Projects   11
   Raw Data Stage Actions: Ingest Data and Create Metadata   12
      Ingesting Known and Unknown Data   12
      Creating Metadata   14
   Refined Data Stage Actions: Create Canonical Data and Conduct Ad Hoc
      Analyses   23
      Designing Refined Data   24
      Refined Stage Analytical Actions   26
   Production Data Stage Actions: Create Production Data and Build Automated
      Systems   28
      Creating Optimized Data   29
      Designing Regular Reports and Automated Products/Services   29
   Data Wrangling within the Workflow Framework   30

3. The Dynamics of Data Wrangling. 31
   Data Wrangling Dynamics   31
      Additional Aspects: Subsetting and Sampling   32
   Core Transformation and Profiling Actions   34
   Data Wrangling in the Workflow Framework   36
      Ingesting Data   36

Describing Data   37
Assessing Data Utility   37
Designing and Building Refined Data   37
Ad Hoc Reporting   38
Exploratory Modeling and Forecasting   39
Building an Optimized Dataset   39
Regular Reporting and Building Data-Driven Products and Services   40

4. Profiling. 43
Overview of Profiling   43
Individual Value Profiling: Syntactic Profiling   44
Individual Value Profiling: Semantic Profiling   44
Set-Based Profiling   45
Profiling Individual Values in the Candidate Master File   46
  Syntactic Profiling in the Candidate Master File   47
  Set-Based Profiling in the Candidate Master File   48

5. Transformation: Structuring. 51
Overview of Structuring   51
Intrarecord Structuring: Extracting Values   52
  Positional Extraction   52
  Pattern Extraction   54
  Complex Structure Extraction   55
Intrarecord Structuring: Combining Multiple Record Fields   56
Interrecord Structuring: Filtering Records and Fields   57
Interrecord Structuring: Aggregations and Pivots   57
  Simple Aggregations   58
  Column-to-Row Pivots   59
  Row-to-Column Pivots   59

6. Transformation: Enriching. 61
Unions   61
Joins   62
Inserting Metadata   63
Derivation of Values   63
  Generic   63
  Proprietary   64

7. Using Transformation to Clean Data. 67
Addressing Missing/NULL Values   67
Addressing Invalid Values   67

8. Roles and Responsibilities. 69

Skills and Responsibilities   69
  Data Engineer   70
  Data Architect   71
  Data Scientist   71
  Analyst   72
Roles Across the Data Workflow Framework   73
Organizational Best Practices   74

9. Data Wrangling Tools. 77

Data Size and Infrastructure   78
Data Structures   78
  Excel   79
  SQL   79
  Trifacta Wrangler   79
Transformation Paradigms   79
  Excel   80
  SQL   80
  Trifacta Wrangler   81
Choosing a Data Wrangling Tool   82

Foreword

Through the last decades of the twentieth century and into the twenty-first, data was largely a medium for bottom-line accounting: making sure that the books were balanced, the rules were followed, and the right numbers could be rolled up for executive decision-making. It was an era focused on a select group of IT staff engineering the "golden master" of organizational data; an era in which mantras like "garbage in, garbage out" captured the attitude that only carefully engineered data was useful.

Attitudes toward data have changed radically in the past decade, as new people, processes, and technologies have come forward to define the hallmarks of a *data-driven organization*. In this context, data is a medium for top-line value generation, providing evidence and content for the design of new products, new processes, and ever-more efficient operation. Today's data-driven organizations have analysts working broadly across departments to find methods to use data creatively. It is an era in which new mantras like "extracting signal from the noise" capture a different attitude of agile experimentation and exploitation of large, diverse sources of data.

Of course, accounting still needs to get done in the twenty-first century, and the need remains to curate select datasets. But the data sources and processes for accountancy are relatively small and slow to change. The data that drives creative and exploratory analyses represents an (exponentially!) growing fraction of the data in most organizations, driving widespread rethinking of processes for data and computing—including the way that IT organizations approach their traditional tasks.

The phrase *data wrangling*, born in the modern context of agile analytics, is meant to describe the lion's share of the time people spend working with data. There is a common misperception that data analysis is mostly a process of running statistical algorithms on high-performance data engines. In practice, this is just the final step of a longer and more complex process; 50 to 80 percent of an analyst's time is spent wrangling data to get it to the point at which this kind of analysis is possible. Not only does data wrangling consume most of an analyst's workday, it also represents much of the analyst's professional process: it captures activities like understanding what data is available; choosing what data to use and at what level of detail; understanding how to meaningfully combine multiple sources of data; and deciding how to distill the results to a size and shape that can drive downstream analysis. These activities represent the hard work that goes into both traditional data "curation" and modern data analysis. And in the context of agile analytics, these activities also capture the creative and scientific intuition of the analyst, which can dictate different decisions for each use case and data source.

We have been working on these issues with data-centric folks of various stripes— from the IT professionals who fuel data infrastructure in large organizations, to professional data analysts, to data-savvy "enthusiasts" in roles from marketing to journalism to science and social causes. Much is changing across the board here. This book is our effort to wrangle the lessons we have learned in this context into a coherent overview, with a specific focus on the more recent and quickly growing agile analytic processes in data-driven organizations. Hopefully, some of these lessons will help to clarify the importance—and yes, the satisfaction—of data wrangling done well.

CHAPTER 1

Introduction

Let's begin with the most important question: why should you read this book? The answer is simple: you want more value from your data. To put a little more meat on that statement, our objective in writing this book is to help the variety of people who manage the analysis or application of data in their organizations. The data might or might not be "yours," in the strict sense of ownership. But the pains in extracting value from this data are.

We're focused on two kinds of readers. First are people who manage the analysis and application of data indirectly—the managers of teams or directors of data projects. Second are people who work with data directly—the analysts, engineers, architects, statisticians, and scientists.

If you're reading this book, you're interested in extracting value from data. We can categorize this value into two types along a temporal dimension: near-term value and long-term value. In the near term, you likely have a sizable list of questions that you want to answer using your data. Some of these questions might be vague; for example, "Are people really shifting toward interacting with us through their mobile devices?" Other questions might be more specific: "When will our customers' interactions primarily originate from mobile devices instead of from desktops or laptops?"

What is stopping you from answering these questions? The most common answer we hear is "time." You know the questions, you know how to answer them, but you just don't have enough hours in the day to wrangle your data into the right form.

Beyond the list of known questions related to the near-term value of your data is the optimism that your data has greater potential long-term value. Can you use it to forecast important seasonal changes? What about risks in your supply chain due to weather or geopolitical shifts? Can you understand how the move to mobile is affecting your customers' purchasing patterns? Organizations generally hire data scientists to take on these longer-term, exploratory analyses. But even if you have the requisite skills to tackle these kinds of analyses, you might still struggle to be allocated sufficient time and resources. After all, exploratory analytics projects can take months, and often contain a nontrivial risk of producing primarily negative or ambiguous results.

As we've seen, the primary impediment to realizing both the short-term and long-term value of your data is time: your limited time and your organization's limited time. In this book, we describe how improving your data wrangling efforts can create the time required to get more near-term and long-term value from your data. In Chapters 1-3, we describe a workflow framework that links activities focused on both kinds of value, and explain how data wrangling factors into those activities and into the overall workflow framework. We introduce the basic building blocks for a data wrangling project: data flow, data wrangling activities, roles, and responsibilities. These are all elements that you will want to consider, at a high level, when embarking on a project that involves data wrangling. Our goal is to provide some helpful guidance and tips on how to coordinate your data wrangling efforts, both across multiple projects by making sure your wrangling efforts are constructive as opposed to redundant or conflicting, and within a single project by taking advantage of some standard language and operations to increase productivity and consistency.

There's more to effective data wrangling than just clearly defined workflows and processes; to most effectively wrangle your data, you should also understand which transformation actions constitute data wrangling, and, most important, how you can use those transformations to produce the best datasets for your analytic activities.

Those nitty-gritty transformations constitute our discussion in Chapters 4-7. You can think of those chapters as a rough "how-to" guide for data wrangling. That said, we do not intend this book to provide a comprehensive tutorial on all possible data wrangling methods. Instead, we want to give you a collection of techniques that you can use when moving through the stages of the data workflow framework.

As we introduce each of the key transformation and profiling activities that comprise data wrangling, we will walk through a theoretical data project involving a publicly available dataset containing US campaign finance information. You can walk through the project along with us in your data wrangling tool of choice.

Finally, we end by discussing roles and responsibilities in a data wrangling project in Chapter 8, and exploring a selection of data wrangling tools in Chapter 9.

Throughout the book, we ground our discussion in example data, transformations of that data, and various visual and statistical views of that data. Along those lines, we open with a story about Facebook.

Magic Thresholds, PYMK, and User Growth at Facebook

Growth is about tapping and delivering value to the yet unserved part of your market. Facebook stands as a quintessential example of how to drive growth. Toward the end of 2015, Facebook reported more than one billion daily active users with a year-over-year growth around 17 percent.[1] There are, of course, many factors that have contributed to this growth. We'll focus here on a series of data-driven insights that armed Facebook with strategies to deliver robust growth, year over year over year.

Growth is ultimately about increasing the number of actively engaged users and customers. It follows a simple equation:[2]

active users = new users + returning users + resurrected users

A critical aspect of growth is bringing new users and customers to your product or service. But just as critical is delivering value to new users so that they stay engaged. Ideally, users are "returning" (i.e., active from one period to the next). However, depending on how you are tracking engagement, you might see blips of inactivity followed by reengagement (placing these users in the "resurrected" group in the aforementioned equation). We'll focus on this second critical aspect of growth—delivering value to new users quickly so that they are motivated to stay engaged.

As Alex Schultz, vice president of growth at Facebook, points out, the primary value for Facebook users revolves around connecting people to the content from their friends.[3] Obviously, for this to work, users need friends on Facebook. But is this the only thing that matters—any content from any friend? Common sense would tell you that this can't be true, and that people engage with some content more than other content. So here we have a set of near-term questions to answer:

- How many friends does a new Facebook user need to be *X*-percent likely to return as a user in 30 days? In 60 days? In 180 days?

- For new users, what characteristics of their friends stand out to differentiate between new users who churn (leave the platform and don't come back) versus those who remain active?

- Do the preceding findings change by user cohort (groups of users that initially joined Facebook at around the same time)?

Answering questions like these is the purview of the Growth and Analytics team at Facebook. Interestingly, the team found a magic threshold that captured a key predic-

---

1 *https://s21.q4cdn.com/399680738/files/doc_financials/annual_reports/2015-Annual-Report.pdf*

2 *https://medium.com/swlh/diligence-at-social-capital-part-1-accounting-for-user-growth-4a8a449fddfc#.w7lptg3n4*

3 *https://blog.kissmetrics.com/alex-schultz-growth/*

tor of long-term user engagement: new users should connect to 10 friends within 14 days. Magic thresholds have two key characteristics: first, they should correspond to a concise Key Performance Indicator (KPI) target that predicts (and if you are lucky, drives) the impact you want; and second, they should be actionable. KPI targets are standard across industries and departments, but what sets a magic threshold apart is that it exposes the core dynamic of the system and provides a lever for achieving a desired outcome. In the case of Facebook, connecting to friends quickly is a critical driver of value for new users, and if Facebook can find ways to reach that threshold for more new users, more new users should stay engaged over the long term.

This magic threshold has the advantage of encoding the core value proposition of Facebook: users connecting to their friends. It also has the advantage of coordinating a number of product decisions to help satisfy this threshold for as many new users as possible.

So, how does Facebook find friends for new users? There are simple, manual mechanisms that allow new users to import their email contact lists (which Facebook then triangulates with its known list of users). This provides short-term value. Facebook also utilizes more sophisticated mechanisms to link users to friends. We consider these mechanisms to fall into the realm of long-term value, in part because the depth of analyses and experimentation that are required to robustly expose this value take months to years. But more importantly, these in-depth analyses give rise to data-driven services that automatically perform the desired operations.

In Facebook's case, one of the core systems used to drive growth, by helping new users connect to friends within Facebook, is known as PYMK, or People You May Know. PYMK is a recommender system, not unlike Amazon's product recommendation system or Netflix's movie/show recommendation system. It employs a well-known and often-used user experience rule: recognition is better than recall. In other words, it's easier and more enjoyable for users to say "yes" or "no" to a series of suggestions than it is for them to generate the content of the suggestions through search or a menu-driven builder experience.

PYMK uses a number of features about the new users and, more important, about the first few friends to whom they have connected. In its most basic form, you can think of PYMK as collecting all the friends of a user's friends to whom they are not currently connected. Then, based on metrics like the number of mutual friends, age similarity, education similarity, and so on, it ranks this list and presents it back to the user as recommendations.

So, with a little bootstrapping from an important contact list or a few manual friend searches, new users on Facebook begin receiving recommendations on who to connect with. The PYMK system that enables these connections has been critical to Facebook's continuous growth.

But the story becomes even more interesting. After some long-running analyses and experimentation, Facebook found that a more effective use of PYMK for user growth was not to focus on recommendations for new users (because bootstrapping is difficult and the early recommendations can come with low-confidence scores), but rather to focus on recommendations to heavy, long-time users of Facebook with vast and diverse connections. Specifically, the key is to recommend new users to the heavy Facebook users. This primes a new user with all sorts of interesting content and the friend network of the heavy user can provide better estimates on friend recommendations directed to the new user.

Although certainly unique in many ways, Facebook's use of data stands as a repeatable process that many other organizations can follow. Starting with a clear motivation— driving user growth—a number of explicit, near-term questions can provide critical insights to improve the business. Over the long term, these insights can blossom into data services that automate and optimize the earlier insights for deeper and additional value.

In Chapter 2, we describe our workflow framework that links near-term and long-term value from data with the variety of activities involved in working with data.

A Data Workflow Framework

In this chapter, we present a framework for working with data. Our goal is to cover the most common sequences of actions that people take as they move through the process of accessing, transforming, and using their data. We'll begin at the end of this process, and discuss the value you will get from your data.

In the introduction, we talked about near-term and long-term value. Another dimension of value to consider is how that value will be delivered into your organization. Will value be delivered directly, through systems that can take automated actions based on data as it is processed? Or will value be delivered indirectly, by empowering people in your organization to take a different course of action than they otherwise would have?

Indirect value
> Data provides value to your organization by influencing people's decisions or inspiring changes in processes. Example: risk modeling in the insurance industry.

Direct value
> Data provides value to your organization by feeding automated systems. Example: Netflix's recommendation system.

Indirect value from data has a long tradition. Entire professions are built on it: accounting, risk modeling in insurance, experimental design in medical research, and intelligence analytics. On a smaller scale, you might have used data to generate reports or interactive visualizations. These reports and visualizations both use data to deliver indirect value. How? When others view your report or visualization, they incorporate the presented information into their understanding of the world and then use their updated understanding to improve their actions. In other words, the data shown in your reports and visualizations indirectly influences other people's decisions. Most of the near-term, known potential value in your data will be delivered indirectly.

Direct value from data involves handing decisions to data-driven systems for speed, accuracy, or customization. The most common example of this involves automatic delivery and routing of resources. In the world of high-frequency trading and modern finance, this resource is primarily money. In some industries, like consumer packaged goods (think Walmart or Amazon), physical goods are routed automatically. A close cousin to routing physical goods is routing virtual ones: digital media companies like Netflix and Comcast use automated pipelines to optimize the delivery of digital content to their customers. At a smaller scale, systems like antilock brakes in cars use data from sensors to route energy to different wheels. Modern testing systems, like the GRE graduate school entrance exam, now dynamically sequence questions based on the tester's evolving performance. In all of these examples, a significant number of operational decisions are directly controlled by data-driven systems without any human input.

How Data Flows During and Across Projects

Deriving indirect, human-mediated value from your data is a prerequisite to deriving direct, automated value. At the outset, human oversight is required to discover what is "in" your data and to assess whether the quality of your data is sufficiently high to use it in direct and automated ways. You can't send data blindly into an automated system and expect valuable results. Reports must be authored and digested to understand the wider potential of your data. As that wider potential comes into focus, automated systems can be designed to use the data directly.

This is the natural progression of data projects: from near-term answering of known questions, to longer-term analyses that assess the core quality and potential applications of a dataset, and finally to production systems that use data in an automated way. Underlying this progression is the movement of data through three main data stages: raw, refined, and production. Table 2-1 provides an overview of this progression. For each stage, we list the primary objectives.

Table 2-1. Data moves through stages

	Data Stage		
	Raw	**Refined**	**Production**
Primary Objectives	• Ingest data • Data discovery and metadata creation	• Create canonical data for widespread consumption • Conduct analyses, modeling, and forecasting	• Create production-quality data • Build regular reporting and automated data products/ services

In the raw stage, the primary goal is to discover the data. When examining raw data, you ask questions aimed at understanding what your data looks like. For example:

- What kinds of records are in the data?
- How are the record fields encoded?
- How does the data relate to your organization, to the kinds of operations you have, and to the other data you are already using?

Armed with an understanding of the data, you can then refine the data for deeper exploration by removing unusable parts of the data, reshaping poorly formatted elements, and establishing relationships between multiple datasets. Assessing potential data quality issues is also frequently a concern during the refined stage, because quality issues might negatively affect any automated use of the data downstream.

Finally, after you understand the data's quality and potential applications in automated systems, you can move the data to the production stage. At this point, production-quality data can feed automated products and services, or enter previously established pipelines that drive regular reporting and analytics activities.

A minority of data projects will end in the raw or production stages. The majority will end in the refined stage. Projects ending in the refined stage will add indirect value by delivering insights and models that drive better decisions. In some cases, these projects might last multiple years. Google's Project Oxygen is a great example of a project that ended in the refined stage.[1] Realizing that managing people is a critical skill for a successful organization, Google kicked off a multiyear study to assess the characteristics of a good manager and then test how effective they could be at teaching those characteristics. The results of the study indireclty influenced employee behavior, but the study data itself was not incorporated into a production pipeline.

The hand-off between IT shared services organizations and lines of business traditionally occurs in the refined stage. In such an environment, IT is responsible for Extract-Transform-Load (ETL) operations. ETL moves data through the three data stages in a centrally controlled manner. Lines of business own the data analysis process, including everything from reporting and ad hoc research tasks, to advanced modeling and forecasting, to data-driven operational changes. This division of concerns and responsibilities has two intended benefits: basic data governance due to centralized data processing, and efficiency gains due to IT engineers reusing broadly useful data transformations.

However, in practice, the perceived benefits of centrally transforming data are often eclipsed by the reality of organizational inefficiencies and bottlenecks. Most of these

---

1 *https://hbr.org/2013/12/how-google-sold-its-engineers-on-management*

bottlenecks arise from line-of-business analysts being dependent upon IT. In the age of agile analytics and data-driven services, there is increasing pressure to speed up the extraction of value from your data. Unsurprisingly, the best plan of attack involves identifying and removing bottlenecks.

In our experience, there are two primary bottlenecks. The first bottleneck is the time it takes to wrangle your data. Even when you start from refined data, there are often nontrivial transformations required to prepare your data for analysis. These transformations can include removing unnecessary records, joining in additional information, aggregating data, or pivoting datasets. We will discuss each of these common transformation actions in more detail in later chapters.

The second bottleneck is the simple capacity mismatch that arises when a large pool of analysts relies on a small pool of IT professionals to prepare "refined" data for them. Removing this bottleneck is more of an organizational challenge than anything else, and it involves expanding the range of users who have access to raw data and providing them with the requisite training and skills.

To help motivate these organizational changes, let's step back and consider the gross mechanics of successfully using data. The most valuable uses of your data will be production uses that take the form of automated reports or data-driven services and products. But every production use of your data depends on hundreds or even thousands of exploratory, ad hoc analyses. In other words, there is a funnel of effort leading to direct, production value that begins with exploratory analytics. And, as with any funnel, your conversation rate will not be 100 percent. You'll need as many people as possible exploring your data and deriving insights in order to discover a relatively small number of valuable applications of your data.

As Figure 2-1 demonstrates, a large number of raw data sources and exploratory analyses are required to produce a single valuable application of your data.

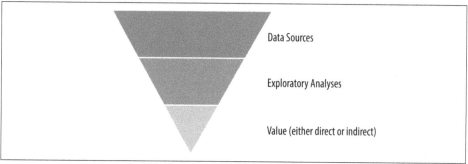

Figure 2-1. Data value funnel

When it comes to delivering production value from your data, there are two critical points to consider. First, data can produce insights that are not useful to you and your business. These insights might not be actionable, or their potential impact might be too small to warrant a change in existing processes. A good strategy for mitigating this risk is to empower the people who know your business priorities to explore your data. Second, your exploratory analytics efforts should be as efficient as possible. This brings us back to data wrangling. The faster you can wrangle data, the more explorations of your data you can conduct, and the more analyses you will be able to move into production. Ultimately, implementing an effective data wrangling workflow can enable more business analysts to explore a larger quantity of data at a faster pace.

Connecting Analytic Actions to Data Movement: A Holistic Workflow Framework for Data Projects

We began this chapter with a discussion of the direct and indirect value delivered by data projects.

In this section, we expand our discussion of data stages into a complete framework that captures the basic analytic actions involved in most data projects. Figure 2-2 illustrates the overall framework and will serve as our map through the rest of the book.

As Figure 2-2 illustrates, data moves through stages, from raw to refined to production. Each stage has a small set of primary actions. The actions come in two types: in the top three boxes in Figure 2-2 are actions whose results are the data itself, and in the bottom six boxes are actions whose results are derived from or built on top of the data inferences (e.g., insights, reports, products, or services). For simplicity, the connecting links between actions in Figure 2-2 are drawn in one direction. However, real data projects will often loop back through actions, iterating toward better results.

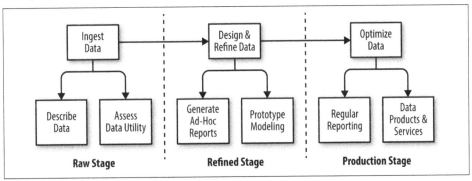

Figure 2-2. A holistic workflow framework for data projects

Of course, many individuals and organizations will customize the steps in this framework to fit their specific needs. Although we describe each possible action in our framework, not every data project will involve all of these actions. You might decide to define variants of each action that are tailored to specific customers or business objectives. You might also decide to create multiple locations for refined data and multiple locations for production data. We have seen this frequently at organizations where data security is important, and different business units are not allowed to access each other's data. However, most organizations that we have worked with follow the uncustomized version of this framework.

In the rest of this chapter, we'll discuss the actions in Figure 2-2. The discussion will move through the three data stages in order.

Raw Data Stage Actions: Ingest Data and Create Metadata

There are three primary actions in the raw data stage: ingestion of data, creation of generic metadata, and creation of propriety metadata. We can separate these actions into two groups based on their output, as shown in Figure 2-3. One group is focused on outputting data—the two ingestion actions. The second group is focused on outputting insights and information derived from the data—the metadata creation actions.

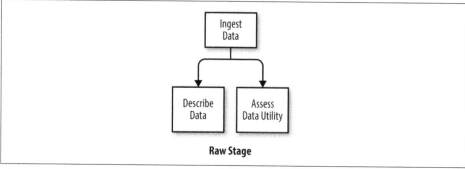

Figure 2-3. Primary action and output actions in the raw data stage

Ingesting Known and Unknown Data

The process of ingesting data can vary widely in its complexity. At the less complex end of the spectrum, many people receive their data as files via channels like email, shared network folders, or FTP websites. At the more complex end of the spectrum, modern open source tools like Sqoop, Flume, and Kafka enable more granular and real-time transferring of data, though at the cost of requiring nontrivial software engineering to set up and maintain. Somewhere in the middle of this spectrum are proprietary platforms like Alteryx, Talend, and Informatica Cloud that support a variety of data transfer and ingestion functionality, with an eye toward easing of configuration and maintenance for nonengineers.

In traditional enterprise data warehouses, the ingestion process involves some initial data transformation operations. These transformations are primarily aimed at mapping inbound elements to those elements' standard representations in the data warehouse. For example, you might be ingesting a comma-separated values (CSV) file and need each field in that file to correspond to a particular column in a relational data warehouse. After it is transformed to match the syntax rules defined by the warehouse, the data is stored in predefined locations. Often this involves appending newly arrived data to related prior data. In some cases, appends can be simple, literally just adding new records at the "end" of the dataset. In other cases, when the incoming data contains edits to prior data as well as new data, the append operation becomes more complicated. These scenarios often require you to ingest new data into a separate location, where more complex merging rules can be applied during the refined data stage.

Some modern NoSQL databases like MongoDB or Cassandra support less-rigid syntax constraints on incoming data while still supporting many of the classic data access controls of more traditional warehouses. Further along the spectrum (toward relaxed constraints on incoming data) are basic storage infrastructures like HDFS and Amazon S3 buckets. For most users, S3 and HDFS look and act like regular filesystems. There are folders and files. You can add to, modify, and move them around. And, if necessary, you can control access on a per-file, per-user basis.

The primary benefit of modern distributed filesystems like HDFS and S3 is that data ingestion can be as simple as copying files or storing a stream of data into one or more files. In this environment, the work to make this data usable and accessible is often deferred until the data is transformed and moved to the refined data stage. This style of data ingestion is often referred to as *schema-on-read*. In schema-on-read ingestion, you do not need to construct or enforce a usable data structure until you need to use the data. Traditional data warehouses, in contrast, require *schema-on-write*, in which the data must adhere to certain structural and syntactic constraints in order to be ingested.

In other words, the two ends of the ingestion complexity spectrum differ based on when the initial enforcement of data structure happens. However, it is important to note that along this entire spectrum of ingestion infrastructures, you will still require a separate refined data stage. This is because refined data has been further wrangled to align with foreseeable analyses.

Let's consider an example data ingestion use case. It is common practice for consumer packaged goods (CPG) retailers (e.g., Walmart and Target) and manufacturers (e.g., Pepsico and General Mills) to share data about their supply chains. This data enables better forecasting, helping both sides to better manage inventory. Depending on the size of the companies, data might be shared on a daily, weekly, or monthly basis. The ingestion complexity comes from the many-to-many partnerships in this ecosystem:

retailers sell products from many manufacturers, and manufacturers sell products to many retailers. Each of these companies produces data in different formats and conforms to different syntactic conventions. For example, each company might refer to products by using their own product IDs or product descriptions. Or some companies might report their data at case or bundle levels instead of the individual units that an end consumer would purchase. Retailers with strong weekly patterns (e.g., much higher sales activity on weekends versus weekdays) might report their overall sales activity on a weekly basis instead of a monthly basis. Even retailers that report their data at the same frequency might define the beginning and ending of each period differently. Further complexity arises in retailer sales data when consumers return purchased goods. Return transactions require amendments to previously shared sales data, often going back multiple weeks.

The ingestion processes for these CPG companies can range from simple file transfers, which wait for the refined data stage to tackle the potentially complex wrangling tasks required to sort out the aforementioned difficulties, to more engineered ETL processes that fix some of these difficulties as the data is ingested. In either case, both retailers and manufacturers are interested in forecasting future sales. Because these forecasts are regularly refreshed, and because the historical sales data on which they are based can and is amended, most large CPG companies work with supply chain data in a time-versioned way. This means that a forecast for the first week of January 2017 based on data received through August 31, 2016 is kept separate and distinct from a forecast for the same first week of January 2017 using data received through September 30, 2016.

In addition to storing data in time-versioned partitions, data from different partners is often ingested into separate datasets. This greatly simplifies the ingestion logic. After ingestion, as the data moves into the refined stage, the separate partner datasets are harmonized to a standard data format so that cross-partner analyses can be efficiently conducted.

Creating Metadata

In most cases, the data that you are ingesting during the raw data stage is known; that is, you know what you are going to get and how to work with it. But what happens when your organization adds a new data source? In other words, what do you do when your data is partially or completely unknown? Ingesting unknown data triggers two additional actions, both related to the creation of metadata. One action is focused on understanding the characteristics of your data, or describing your data. We refer to this action as generating generic metadata. A second action is focused on using the characteristics of your data to make a determination about your data's value. This action involves creating custom metadata.

Before discussing the two metadata-producing actions, let's cover some basics. Datasets are composed of records. Records are composed of fields. Records often represent or correspond to people, objects, relationships, or events. The fields within a record represent or correspond to measurable aspects of the person, object, relationship, or event. For example, if we consider a dataset that contains transactions from a store, each record could correspond to a single purchase, and fields might represent the monetary value of the purchase, the specific goods purchased, the time of the purchase, and so on. If you're used to working with relational data, you might be used to speaking about rows and columns. Records are synonymous with rows, and fields are synonymous with columns.

When you are describing your data, you should be focused on understanding the structure, granularity, accuracy, temporality, and scope of your data. Structure, granularity, accuracy, time, and scope are key aspects of representational consistency. As such, they are also the characteristics of a dataset that must be tuned or improved by your wrangling efforts.

Beyond generic metadata descriptions, the data discovery process often requires inferring and creating custom metadata related to the potential value of your data. Whereas the generic metadata should be broadly useful to anyone working with the dataset, custom metadata contextualizes this information to a specific analysis or organization. In other words, custom metadata builds on or extends generic metadata.

Both generic and custom metadata are composed of the same base set of characteristics: structure, granularity, accuracy, temporality, and scope. We will discuss each of these characteristics in turn and explain how you can better understand them in the context of your data.

Structure

The structure of a dataset refers to the format and encoding of its records and fields. We can consider datasets on a spectrum related to the homogeneity of their records and fields. At one end of the spectrum, the dataset is "rectangular" and can be formatted as a table with a fixed number of rows and columns. In this format, rows in the table correspond to records, and columns correspond to fields.

If the record fields in a dataset are not consistent (some records have additional fields, others are missing fields, etc.), you could be dealing with a "jagged" table. Such a table is no longer perfectly rectangular. Data formats like JSON and XML support datasets like this, in which record fields are not fully consistent.

Further along the spectrum are datasets with a heterogeneous set of records. For example, a heterogeneous dataset from a retail organization might mix customer information and customer transactions. This is a common occurrence if you consider the tabs in a complex Excel workbook. Most analysis and visualization tools will require these different kinds of records to be split into separate files or tables. Analyses that require all of these records will often start by joining or blending the records together based on shared fields.

The encoding of a dataset specifies how the record fields are stored and presented to the user. Are the timestamps in local time zones or mapped to UTC? Do date/time fields conform to specifications like ISO 8601? Are ages stored in years, days, or fractions of decades? Are account balances in US dollars? Are the postal codes up to date? Are accented characters supported, or have they been replaced with their nonaccented counterparts? Are right-to-left writing systems, like Arabic, supported?

In many cases, it is advisable to encode a dataset in plain text. This makes the file human-readable. A major drawback to encoding the file in plain text is the size of the file; it is far more space efficient to use binary encodings of numerical values, or to compress the file using standard algorithms like gzip and bzip.

Assessing the structure of your dataset is primarily a generic metadata question. Before you can begin wrangling your data, you need to understand how that data is structured. This often requires counting the number of records and fields in a dataset, and determining the dataset's encoding.

Beyond these generic metadata concerns, you might need to generate custom metadata pertaining to the specific structure of your dataset. Earlier in this chapter, we discussed CPG organizations that need to work with data provided by their external trading partners. Because this data originates outside of the organization, CPG analysts often need to determine the custom fields that each trading partner is adding to the dataset. Maybe Walmart's datasets include a slightly different set of fields than those for Target. When you make these determinations, you generate custom metadata.

Basic Questions to Assess Structure

Here are some of the questions you need to ask when assessing data structure:

- Do all records in the dataset contain the same fields?
- How can you access the same fields across records? By position? By name?
- How are the records delimited/separated in the dataset? Do you need sophisticated parsing logic to separate the records from one another?
- How are the record fields delimited from one another? Do you need to parse them?
- How are record fields encoded? Human readable strings? Binary numbers? Hash keys? Compressed? Enumerated codes?

- What is the complexity of the encoding? Primitive elements like integers, decimal numbers, short strings, and so on? Higher-order elements like key-value sets or arrays?
- What are the semantics of the encoded data? Do these semantics entail associated data quality, consistency, and accuracy checks?
- What are the relationship types between records and the record fields? Singular (record should have one and only one value for a field, like customer date of birth)? Set-based (record could have many values for the field, like customer shipping addresses)?

Granularity

The granularity of a dataset refers to the kinds of entities that each data record represents or contains information about. In their most common form, records in a dataset will contain information about many instances of the same kind of entity.

We typically describe granularity in terms of coarseness and fineness. In the context of data, this means the level of depth or the number of distinct entities represented by a single record of your dataset. For example, a dataset in which a single record represents a single sales transaction by a single customer at a particular store would have a fine granularity. A dataset in which each record represents the total sales in each store for each day would have a coarse granularity. At an even coarser granularity, you might have a dataset in which each record represents total sales by region and week. Depending on your intended use of the dataset, the granularity might be just right, too coarse, or too fine.

There are subtleties to assessing the granularity of a dataset that involve applying organizational knowledge. These are examples of creating custom metadata related to granularity. For example, at first glance, a dataset might appear to contain records that represent customers. However, those records might in fact correspond to all known contacts of your company (only a subset of whom are actual, paying customers). Moreover, this contacts dataset might contain multiple entries for the same person, resulting from that person signing up to receive information through multiple channels (e.g., Facebook and via a direct visit to your website). In this case, a more appropriate description of the granularity of the dataset might be "registration events."

Basic Questions to Assess Data Granularity

Following are questions that you need to ask when assessing data granularity:

- What kind of thing (person, object, relationship, event, etc.) do the records represent?

- Are the records homogeneous (represent the same kinds of things)? Or heterogeneous?

- What alternative interpretations of the records are there? For example, if the records appear to be customers, could they actually be all known contacts (only some of which are customers)?

Accuracy

The accuracy of a dataset refers to its quality. In other words, the values populating record fields in the dataset should be consistent and accurate. For example, consider a customer actions dataset. This dataset contains records corresponding to when customers added items to their shopping carts. In some cases, the reference to the item added to the cart in the record is not accurate; perhaps a UPC code is used but some of the digits are missing, or the UPC code is out of date and has since been reused for a different item. Inaccuracies would, of course, render any analysis on the dataset problematic.

Other common inaccuracies are misspellings of categorical variables, like street or company names; lack of appropriate categories, like ethnicity labels for multiethnic people; underflow and overflow of numerical values; and missing field components, like a timestamp encoded in a 12-hour format but missing an AM/PM indication.

Accuracy assessments can also apply to frequency outliers—values that occur more or less often than you'd expect. Making frequency assessments is primarily a custom metadata concern, given that determining whether the range of values in a dataset is accurate relies on organizational knowledge. Returning to our CPG example, a supply-chain analyst might know that a certain trading partner only reports UPCs within a particular range. In this case, the analyst needs to generate custom metadata around the accuracy of the UPC code distribution in a dataset.

Basic Questions to Assess Data Accuracy

When assessing data accuracy, you need to look at the following:

- Many accuracy issues are type specific:

  — For date times, are time zones included or adjusted into a standard time zone like UTC? If the times are presented in 12-hour format, is AM/PM demarcated? Are the positions of month and day fields ambiguous (e.g., 01/02/17 versus 02/01/17)? Are there signs that the information is wrong (e.g., across your customer dataset, do you have an unusual amount of people with date-of-births corresponding to the first year of a decade like 1970, 1980, 1990, or born in January or on the first date of the month)?

  — For addresses, are the address components consistent? Is the correct postal code associated with the address? If there are GPS coordinates, do they match the address?

  — For numeric items like phone numbers and UPC codes, are digits missing or are subcomponents of the number invalid (like an invalid area code on a phone number)?

  — For names, are there misspellings? Or missing fields (like no first name for a customer)?

  — For email addresses, is the email domain valid?

  — For sales transactions, are the currency amounts all in the same currency? Do they show signs of inaccuracy or fraud (i.e., a preponderance of unlikely values)?

- Some accuracy is related to the mechanisms that produce the data:

  — Is there any chance of sensor drift that has caused systematic inaccuracies over time?

  — Is the data entered by people? If so, there might be a high incidence of misspellings and nonstandard abbreviations.

- Finally, you will want to understand the distribution of inaccuracies across your dataset:

  — What is the measurable distribution of inaccuracies?

  — Does the distribution of inaccuracies affect a large number of records?

  — Are inaccuracies concentrated in a particular subset of records?

Temporality

A data record is a representation of an entity at a particular time (or set of times). Accordingly, even though a dataset might have been an accurate and consistent representation at the time it was created, subsequent changes might render the representation inaccurate or inconsistent. For example, you might use a dataset of customer actions to determine the distribution of items people own. Weeks or months after an initial sale, however, some of these items might be returned. Now the original dataset, although an accurate representation of the original sales transaction, is no longer an accurate representation of the items a person owns.

The time-sensitive nature of representations, and hence of datasets, is an important aspect that should be explicitly noted. Sometimes this is done on a per-record basis (e.g., each customer action contains a timestamp). Data records that do not correspond to events are less likely to contain embedded timestamps; for example, a customer database with names, addresses, demographics, or a reference dataset of UPCs with item descriptions. However, even when time is not explicitly represented in a dataset, it is still important to understand how time may have impacted the records in the dataset. In the first example, a customer might move and his address can become inaccurate. In the second example, UPC numbers might be recycled and item descriptions can become inaccurate. In all such cases, knowing when the dataset was generated can provide valuable insight into potential inaccuracies and the appropriate wrangling actions needed to remediate those inaccuracies.

Basic Questions to Assess Data Temporality

The following questions will help you assess data temporality:

- When was the dataset collected?
- Were all the records and record fields collected/measured at the same time? If not, is the temporal range significant?
- Are the timestamps associated with collection of the data known and available (as a record field) or as associated metadata?
- Have some records or record field values been modified after the time of creation? Are the timestamps of these modifications available?
- In what ways can you determine if the data is "stale"? For example, you might have purchased a marketing leads database and want to verify the contact information for the people represented in the dataset. Is it sufficient to sample the records and manually verify the data? Can you automatically verify it by using third-party services?

- If there are conflicting values in the data (e.g., multiple mailing addresses for a person), can you use timestamps to determine which value is "correct"?
- Can you forecast when the values in the dataset might become "stale"?

Scope

The scope of a dataset has two major dimensions. The first dimension concerns the number of distinct attributes represented in a dataset. For example, for each customer action, we might know when it happened (e.g., a timestamp) and some details about it (like which UPC a customer added to a basket). The second dimension concerns the attribute-by-attribute population coverage: are "all" the attributes for each field represented in the dataset, or have some been randomly, intentionally, or systematically excluded?

Let's begin our discussion of the importance of scope by addressing the number of distinct attributes in a dataset. Each distinct attribute is generally represented in a dataset by a single field. A dataset with a wide scope will have a large number of fields, whereas a dataset with a narrow scope will have a small number of fields. You could imagine that a customer information dataset with a very wide scope could have hundreds of distinct fields, each representing a different attribute of a customer (age, salary, ethnicity, family size, etc.).

Obviously, increasing the scope of a dataset by adding in more entity characteristics extends the analytical potential of the dataset. However, as with granularity, you want to include only as much detail as you might use, but no more. The level of detail required might depend on your analytics methodology. Some methodologies, like deep learning, call for keeping many redundant attributes and letting statistical methods boil those many attributes down to a smaller number. Other methodologies operate best with a limited number of attributes.

Regarding population coverage, the most common scenario is that not all of the possible entities have been represented. Sometimes, missing records are the result of external factors, like the logging infrastructure failing due to a power outage. Other times you're missing some of the data due to an operational or logistical error: a chunk of a file was destroyed, a sensor was knocked off power, somebody didn't give you an entire file, or somebody redacted the dataset. The cause of the missing data can be helpful to know because it can help account for the bias downstream. For example, if you know that a dataset was scrubbed of people under 18 for legal reasons, you should account for the average age in the dataset being higher than the true population that the dataset (imperfectly) represents.

More generally, we might want to introduce the notion of an idea or "true" source of data in the real world, and an imperfect record of that data that was acquired. When statisticians or scientists speak of "sampling," they usually are talking about this—

acquiring a small but representative subset of the "true" data as a stored dataset. The ideal (true) source of data might be infeasible to capture (think about the temperature of every cubic inch of air at every second hovering over the globe—we sample that to get temperature readings at particular thermometers). The data that is recorded is inherently a sample. This is a big difference between real-world (e.g., IoT) data and electronic data like transactions or web click logs.

It is important to understand any systematic bias in a dataset, because for cases in which systematic bias exists, any analytical inferences made using the biased dataset can be rendered invalid. A canonical example is drug trial analysis, where analysts are concerned with assessing the efficacy of the drug being trialed. The granularity of drug trial datasets is often at the level of patients in the trial. If, however, the scope of the dataset has been intentionally reduced by systematically removing records associated with some patients in the trial (either because they died during the trial or because they began showing abnormal biometrics), there's a good chance that any analysis on the dataset is likely to misrepresent the actual impact of the drug.

The most common custom metadata question is: "How can this dataset blend with (join or union to) my other datasets?" To answer this question, we must understand the scope of the datasets. In some cases, the new dataset might represent an extension of an existing dataset. Sometimes, this extension is disjoint, involving records representing entities in the overall population that are missing in your existing datasets. Sometimes, the dataset extension is overlapping, creating a need to deduplicate or harmonize records representing the same entity. In other cases, a new dataset might provide additional record fields (e.g., household disposable income to be matched against your customer list or counts of different kinds of accidents by postal code).

Basic Questions to Assess Data Scope

The following questions will help you assess the scope of your data:

- Given the granularity of the dataset, what characteristics of the things (e.g., people, objects, relationship, events, etc.) represented by the records are captured by the record fields? What characteristics are not captured?

- Are the record fields consistent? For example, does the customer's age field make sense relative to the date-of-birth field? If the record corresponds to a purchase transaction, does the cost of the listed set of purchased items add up to the total transaction amount?

- For the analysis that you want to perform, can you deduce or infer additional relevant characteristics from the ones that you have? For example, can you infer the demographics of the people in a household from partner and dependents record fields?

- Are the same record fields available for all records? Are they accessible via the same specification (position, name, etc.)?

- Do the records in the dataset represent the entire population of associated things (people, objects, relationships, events, etc.)? Are there missing records (e.g., things in the population, say people, with no associated record)? Are the missing records randomly missing or systematically missing?

- Are there multiple records for the same thing? If so, does this change the granularity of your dataset (e.g., from customers to contacts) or require some amount of deduplication before analysis?

- Does the dataset contain a heterogeneous set of records (representing different kinds of entities)? If so, what is the relationship between the different kinds of records?

Refined Data Stage Actions: Create Canonical Data and Conduct Ad Hoc Analyses

After you have ingested your raw data and fully understood the metadata aspects of your raw data, the next major stage in data projects involves refining the data and conducting a broad set of exploratory analyses. Three primary actions define this stage, as shown in Figure 2-4: design and preparation of "refined" data, ad hoc reporting analyses, and exploratory modeling and forecasting. As in the raw data stage, these actions can be separated into two groups distinguished by their output. One group is focused on outputting refined data that enables immediate application to a wide range of analyses. The second group is focused on outputting insights and information derived from the data, ranging from simple reporting to complex models and forecasts.

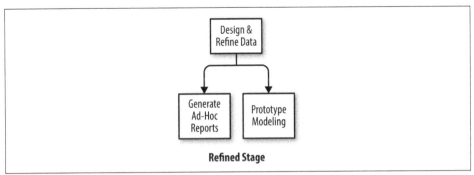

Figure 2-4. Primary action and output actions for the refined data stage

Designing Refined Data

The overarching goal in designing and creating your refined data is to simplify the foreseeable analyses you want to perform. Of course, you won't foresee every analysis. The likely scenario is that insights and patterns gleaned from an initial set of analyses will inspire new analysis directions you hadn't previously considered. To support these new directions, you might need to compile new refined datasets (or, at least, modify the ones you already have). We can, and often do, iterate between the actions in the refined data stage.

In the raw data stage, ingestion involves minimal data transformation—just enough to comply with the syntactic constraints of the data storage system. By contrast, the act of designing and preparing "refined" data often involves a significant amount of transformation. These transformations are often guided by the range of analyses that you plan to conduct on the data, and by the metadata-generating activities undertaken during the raw data stage. If you previously identified quality and consistency issues with the dataset's structure, granularity, accuracy, time, or scope, those issues should be remedied during the refined data stage. We will address each of these metadata-related issues in turn, and discuss how you can design your refined data to resolve or mitigate such issues.

Addressing structural issues

Most visualization and analysis tools expect tabular data, meaning that every record has the same fields instantiated in the same order. Depending on the structure of the raw data, converting data into tabular format can require significant transformations. Furthermore, for modeling and forecasting purposes, you might need to convert categorical data to separate indicator values for category value; for example, you might need to expand a record field encoding gender into multiple fields corresponding to characteristics like "is male" or "is female."

Addressing granularity issues

From a granularity perspective, it is prudent to build refined datasets at the finest resolution of records that you plan to analyze. For example, let's assume that your sales records show a significant increase in the average amount of a sales transaction, and that you can identify a subset of users who form the driver of this shift (this group makes significantly larger purchases relative to other customers, and removing them from the average calculation produces an average sale amount consistent with the recent past). You might want to understand what exactly differentiates these customers from the rest: Are they buying more expensive items? More items than average? Do they shop more often? Do they seem unaware of sales and coupons? Chances are, answering these questions will require the records of actions leading up to the sales transaction; in this case, the records with the finest granularity are most ideal for your analysis.

However, if the majority of the analyses focus on records of a coarser granularity, (e.g., customer segments or demographic groups, and lifetime purchase totals), it also might make sense to store a version of the dataset at this granularity, as well. Retaining multiple versions of the same dataset with different granularities can help streamline downstream analyses based on groups of records.

Addressing accuracy issues

Another key objective in designing and creating refined datasets is to remedy known accuracy issues.

The main strategies used to handle inaccuracies are to: (a) remove records with inaccurate values (provided the inaccuracies can be detected); (b) retain records with inaccurate values but mark them as inaccurate (which still allows some analyses to be conducted over the dataset); or (c) replace the inaccurate values with default or estimated values in a process known as imputation. As an example, for numerically distributed values like the dollar amount of sales transactions, you might determine that extremely large values are inaccurate. These extreme values might be replaced by a maximum value to ensure that aggregate calculations like average and standard deviation are not overly biased.

Conflicting information between fields (e.g., multiple addresses or significant deviation between a date-of-birth field and an age field) or between a field and applicable business logic (e.g., a transaction amount too large to be possible given constraints in the transaction process) are primary accuracy issues that should be addressed during the refined data stage. In some cases, generally when the percent of records with inaccurate values is small and unlikely to be significant, the appropriate remedy is to remove affected records. For many analyses, removing these records will not materially affect the results. In other cases, the best approach might be to reconcile conflicting information; for example, recalculating customer age using date-of-birth and the current date (or the dates of the events you want to analyze).

In many cases, resolving conflicting or inaccurate data fields in your refined data is best done using an explicit reference to time. Consider the multiple addresses problem in a customer dataset. Perhaps each address is (or was) correct, representing the various home addresses a person has had throughout her life. Assigning date ranges to the addresses might resolve the conflicts. Similarly, a transaction amount that violates current business logic might have occurred prior to that logic being enforced, in which case you might want to keep the transaction in the dataset to preserve the integrity of historical analyses.

The most useful notion of "time" can often require some care. For example, there might be a time when an action occurred, and a time when it was recognized. This is particularly common when dealing with bank transactions. In some cases, an abstract version number might be more appropriate than a timestamp. For example, for data generated by software, it might be more important to record the version of the software that was used rather than the time when the software ran. Similarly, in scientific experiments it might be more useful to know the version of a data file that was analyzed rather than the time that the analysis ran. In general, the appropriate choice of time or version often depends on the specifics of an analysis; therefore, it's best to preserve (and document!) all the timestamps and version numbers that are available for a record.

Addressing scope issues

Stepping back from individual record field values, it is also important to design the scope of your refined datasets so that these datasets include the full required set of records and record fields. For example, suppose that your customer data is split into multiple datasets (one containing contact information, another containing transaction summaries, etc.) but most of your analyses involve all of these fields. It might make sense to create a fully blended dataset with all of these fields to streamline your analyses.

Perhaps the most important scope-related challenge is ensuring your refined datasets have understood population coverage. This means that a dataset should accurately express the relationship between the set of things represented by records in the dataset (people, objects, etc.) and the wider population of those things (e.g., all people and all objects).

Ideally, your dataset contains one and only one record representing each member of the population of things you want to analyze, but more likely, your dataset will contain a subset of records from the complete population. If the subset is a true, random subset of the wider population, your dataset can be analytically used to infer insights about the population as a whole. If, however, the subset of represented things in your dataset is not random—that is, it exhibits some kind of bias—you might be restricted on the kinds of analyses you can validly conduct. It's beyond the scope of this book to discuss the statistical implications of bias between your dataset and the population of things it partially represents, but this is an important note to be aware of when designing refined datasets.

Refined Stage Analytical Actions

To complete our discussion of the refined data stage, we'll describe its two core analytical actions: ad hoc reporting analyses, and exploratory modeling and forecasting.

Reporting is the core action for answering specific questions using your data. You can think of business intelligence analytics and dashboarding as specific forms of reporting. These analyses are primarily retrospective—they use historical data to answer questions about the past or present. The answer to those questions might be as simple as a single number or statistic, or as complex as a full report with additional discussion and description of the findings. The nature of the initial query constrains the consumption of the output—there likely won't be an automated system that can consume the output and take direct action as a result. Instead, the results will provide indirect value by informing and influencing people.

There are many types of ad hoc analyses, ranging from straightforward questions that can be answered in a succinct and definitive manner (e.g., how many customers purchased item X last week, or, what were the top three most viewed documents/pages on our website last month?) to open-ended investigations that might last months or years (e.g., identify the key factors driving the customer trend of switching from desktop to mobile devices). A common starting point for ad hoc analyses is an anomaly in a regular report. Perhaps sales spiked up more than expected or there was a dip in transactions from a particular product line or region of stores. If totally unexpected, the anomaly needs to be assessed from a number of different perspectives. Is there a data reporting or data quality problem? If the data is valid (i.e., the anomaly reflects a change in the world and not just in the representation of the world contained in the dataset), can the anomaly be isolated to a subpopulation? What other changes are correlated with the anomaly? Are all of these changes linked via causal dependencies to one another or to a common root change?

Unlike ad hoc analyses, which are primarily retrospective, modeling and forecasting analyses are primarily concerned with the future. These analyses ask, "What do we expect to happen given what we have observed in the past?" In the case of forecasting, the explicit objective is to predict future events: total sales in the next quarter, percent of customer churn next month, likelihoods of each customer renewing their contracts, and so on. In many cases, these predictions are built on models of how the target prediction depends on and relates to other measurable aspects in your dataset. For some analyses, the useful output is not a prediction (or set of predictions), but the underlying model itself.

In most cases, modeling as an explicit activity is an attempt to understand the relevant factors that drive the behavior (whether it is customer behavior, market movements, shifts in the existence or nature of relationships, frequencies or types of events, etc.) that interests you. Even though modeling is a common prerequisite activity for distinguishing correlated factors from causal factors, it is important to note that more causal analysis will also require some amount of carefully designed experimentation; for example, holding some factors constant while perturbing others and/or modifying the underlying system to decouple or realign factors so that you can assess whether those factors truly drive the behavior.

As an example, recall in Chapter 1 that we talked about Facebook's user growth and their reliance on a measurable factor to the Facebook user experience, the number of friends a user has, to drive user retention. Simple analysis of the relationship between number of friends and long-term user retention would show that it is not a causal relationship. There are long-time users of Facebook who have only a few friends, and there are Facebook users who had many friends who have churned from the platform. Nonetheless, these factors are strongly correlated. More important, the product changes that Facebook made to increase the number of friends that users have demonstrably moves the underlying causal factors that do drive user retention.

Production Data Stage Actions: Create Production Data and Build Automated Systems

After you have refined your data and have begun generating valuable insights from that data, you will naturally start to separate out the analyses that need to be regularly refreshed from the ones that were sufficient as one-off analyses. It's one thing to explore and prototype (which is the focus of activities in the refined data stage), but wrapping those initial outputs in a robust, maintainable framework that can automatically direct people and resources is a whole other ballgame. This takes us into the production data stage.

A solid set of initial insights often leads to statements like: "We should track that measure all the time," or "We can use those predictions to expedite shipping of certain orders." The solutions to each of these statements involve "production systems"; that is, systems that operate in a largely automated way and with a well-defined level of robustness. At a minimum, creating production data requires further optimizations to your refined data (Figure 2-5), and then engineering, scheduling, and monitoring the flow of that optimized data into regular reports and data-driven products and services.

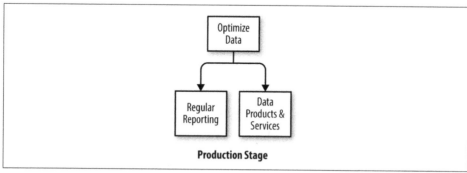

Figure 2-5. Primary action and output actions for the production data stage

Creating Optimized Data

So, what does it mean to optimize your data? In some sense, optimizing data is not unlike designing refined data. You can think of optimized data as the ideal form of your data; it is designed to simplify any additional downstream work to use the data. Unlike refined data, however, the intended use of optimized data should be highly specified.

These specifications go beyond just the scope and accuracy of the values in the data. There are also specifications related to the processing and storage resources that will need to be applied to work with the data on an ongoing basis. These constraints will often dictate the structure of the data, as well as the ways in which that data is made available to the production system. In other words, although the goal of refining data is to support the widest set of analyses as efficiently as possible, the goal of optimizing data is to robustly and efficiently support a very narrow set of analyses.

Designing Regular Reports and Automated Products/Services

Building regular reports and data-driven products and services requires more than just wiring the data into the report generation logic or the service providing logic. One major source of additional work comes from monitoring the flow of data and ensuring that requisite structural, temporal, scoping, and accuracy constraints remain satisfied over time. The fact that data is flowing in these systems implies that new (or updated) data will be processed in an ongoing manner. New data will eventually vary from its historical equivalents (maybe you have updated customer interaction events, or the latest week's sales data). Structural, temporal, scoping, and accuracy constraints define the boundary around permissible variation (e.g., minimum and maximum sales amounts or coordination between record fields like billing address and currency of transaction).

Within the constraints, the reporting and product/service logic must handle the variation. This deviates from exploratory analytics that can, for speed or simplicity, use logic specific to the dataset being analyzed. For production reporting and products/services, the logic must be generalized. Common dataset variations that drive changes to the data wrangling logic include extensions to the value ranges (e.g., current dates or redefining regions or customer segments); new accuracy issues (e.g., previously unseen misspellings); record fields that have been removed or emptied (e.g., for legal compliance purposes, certain information about customers like age or gender might be redacted); appearance of duplicate records; or disappearance of a subset of records (e.g., due to a change in customer segment names, one or more groups might be dropped).

Of course, you could tighten the boundary of permissible variations to exclude things like duplicate records or missing subsets of records. If so, the logic to catch and remedy these variations will likely live in the data optimization action.

Data Wrangling within the Workflow Framework

In this chapter, we have described the characteristic actions and movement of data through the raw, refined, and production stages. Throughout that discussion, we touched on the kinds of data and outputs involved in each stage. But there's still a piece missing from our discussion: what is data wrangling, and how does it relate to our workflow framework?

Fundamentally, data wrangling is the process involved in transforming or preparing data for analysis. If you refer back to our earlier workflow diagrams, data wrangling occurs between the stages; it is the set of actions that allows you to move from raw data to refined data, or from refined data to optimized, production data. Sometimes (particularly in the transition from the raw stage to the refined stage) data wrangling can resemble traditional ETL processes. We consider ETL to be one type of data wrangling, specifically a type of data wrangling managed and overseen by an organization's shared services or IT organization. But data wrangling can also be handled by business users in desktop tools like Excel, or by data scientists in coding languages like Python or R.

We see data wrangling as a core task within every action in the framework. This is not to say, however, that data wrangling tasks will be identical throughout this workflow framework. These wrangling tasks will differ, particularly in the kinds of transformations that are applied. In Chapter 3, we dive into the specifics of data wrangling, describe the various types of data wrangling transformations, and explain how data wrangling tasks differ across our workflow framework.

The Dynamics of Data Wrangling

In Chapter 2, we introduced a framework capturing the variety of actions involved in working with data. Each of these actions involves some amount of data wrangling. In this chapter, we describe the dynamics of data wrangling, the breadth of transformations and profiling required to wrangle data, and how these aspects of data wrangling vary by action in our framework.

Data Wrangling Dynamics

Data wrangling is a generic phrase capturing the range of tasks involved in preparing your data for analysis. Data wrangling begins with *accessing* your data. Sometimes, access is gated on getting appropriate permission and making the corresponding changes in your data infrastructure. Access also involves manipulating the locations and relationships between datasets. This kind of data wrangling involves everything from moving datasets around a folder hierarchy, to replicating datasets across warehouses for easier access, to analyzing differences between similar datasets and assessing overlaps and conflicts.

After you have successfully accessed your data, the bulk of your data wrangling work will involve *transforming* the data itself—manipulating the structure, granularity, accuracy, temporality, and scope of your data to better align with your analysis goals. All of these transformations are best performed with tools that provide meaningful feedback (so that the manipulator is assured that the manipulations were successful). We refer to this feedback as *profiling*. In many cases, a predefined (and, hence, somewhat generic) set of profiling feedback is sufficient to determine whether an applied transformation was successful. In other cases, customized profiling is required to make this determination. In either event, the bulk of data wrangling involves frequent iterations between transforming and profiling your data.

A final set of data wrangling tasks can be understood as *publishing*. Publishing is best understood from the perspective of what is published. In some cases, what is published is a transformed version of the input datasets (e.g., in the design and creation of "refined" datasets). In other cases, the published entity is the transformation logic itself (e.g., as a script that generates the range of statistics and insights in a regular report). A final kind of publishing involves creating profiling metadata about the dataset. These profiling reports are critical for managing automated data services and products.

Figure 3-1 illustrates the simple relationship between these data wrangling steps. As just mentioned, wrangling begins with access. From there, the bulk of time and energy is spent transforming and profiling the results of the transformation. Finally, the desired output is published for downstream consumption. Realistically, data wrangling is far more iterative. In addition to iterations between transforming and profiling the data, there are less frequent iterations that return to accessing data. Likewise, during or soon after publishing a result, you might realize that the output is not exactly correct and you need to apply additional transformations or expose some additional profiling results. These iterations are captured in Figure 3-1.

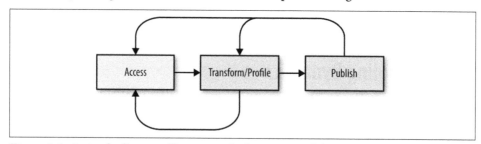

Figure 3-1. A simple diagram illustrating the basic steps of data wrangling

The work begins by obtaining access to your data. You might already have the dataset in a file and need to do little more than double-click to open it. In other cases, you might need to submit a request for access and obtain the necessary credentials. With the data in hand, the bulk of data wrangling involves iterating between applying transformations and assessing the impact of those transformations through profiling. After you have modified the dataset as desired, or authored a robust data transformation script, or produced the profiling statistics and visualizations that showcase aspects of the dataset, any or all of these outputs must be published.

Additional Aspects: Subsetting and Sampling

There are two additional aspects to the dynamics of data wrangling that we believe are vital to finding efficiencies in your data wrangling practice: subsetting data and sampling data. Both are applicable in certain circumstances, versus the general steps and dynamics we discussed earlier, which are broadly applicable.

First, consider the case in which your dataset contains a heterogeneous set of records, differing either in structure (e.g., some records contain more or different fields from the rest) or in granularity (e.g., some records correspond to customers, whereas others correspond to accounts). Faced with this heterogeneity, the best wrangling approach is to split up the original dataset and wrangle each subset separately; then, if necessary, merge the results together again. Process efficiencies are fundamentally rooted in the ability to apply the same processing mechanism (in this case, data transformation logic) to many, similar inputs. At one scale, this amounts to authoring data transformation steps that operate effectively across every record in your dataset. At a wider scale, this amounts to authoring a data transformation script that operates effectively across multiple, similarly structured datasets. This case requires subsetting data for the most efficient data wrangling process.

Now consider the case in which your dataset is too large to manually review each record or when your dataset is so large that even simple transformations require prohibitively long timeframes to complete (or speeding up these transformations would have prohibitively large resource costs); in other words, when you are working with big data. In this case, the iterative process of transforming and profiling your data is materially hampered by the time required to compute and execute transformations.

Suppose that you make a small change to a derived calculation or you change a rule to group a few customer segments into a wider segment. Now you apply this transformation and must wait a minute, or 10 minutes, or half a day to see what the results might look like. Understandably, data wrangling work will dominate your analysis workflows and you won't get through many analyses. The critical approach to speeding up your data wrangling is to work with some samples of the entire dataset that you can transform and profile at interactive time scales (ideally within 100 milliseconds, but occasionally up to a few seconds). Unfortunately, working with samples to speed up your data wrangling is not as straightforward as it sounds.

The complexities of data sampling relative to analysis are well discussed in statistics and surveying texts like Leslie Kish's *Survey Sampling*. Sufficient for our discussion here is to point out the obvious connection between sampling and profiling. Again, our objective is to speed up the task of data wrangling by providing profiling feedback from a sample of the dataset (which can be processed at interactive speeds) versus the entire dataset.

To understand the importance of sampling, consider a simple transformation involving the calculation to determine the length of time each of your customers has been using your product or service based on the date each one registered for it. Chances are, you have some sense of what these ages should be: you started your business 11 years ago, so no customer should show a duration of more than 11 years. You had a big increase in customers about 3 years ago, so you'd expect to see a corresponding bump in the overall age distribution around 3. And so on. These expectations point to a couple of sampling techniques that would be useful for assessing whether your transformation to calculate customer age is working correctly. Namely, you want a sample that contains extreme values (customer records with the earliest and latest registration dates) and that randomly samples over the rest of the records (so that overall distributional trends are visible).

Consider a more complex situation that involves case-based transformations based on record groups; for example, you need to convert transaction amounts to US dollars, and your dataset contains transactions in Euros, GB Pounds, and so on. Each reporting currency requires its own transformation. To assess that all of the currency transformations were applied correctly, you need to profile results covering all of the currencies occurring in your dataset. Samples that cover all groups, or "strata," are often referred to as stratified samples; they provide representation of each group, even though they might bias the overall trends of the full dataset (by overrepresenting small groups relative to large ones, for example). There are numerous techniques for extracting different kinds of samples from large datasets (e.g., see *Synopses for Massive Data: Samples, Histograms, Wavelets, Sketches* by Cormode et al.), and some software packages and databases implement these methods for you.

With an understanding of the basic steps in data wrangling—access, transformation, profiling, and publishing—and how these steps can incorporate aspects of sampling to handle big datasets and split/fix/merge strategies for heterogeneous datasets, we turn our attention now to the core types of transformations and profiling.

Core Transformation and Profiling Actions

The core tasks of data wrangling are transformation and profiling, and the general workflow involves quick iterations (on the order of seconds) between these tasks. Our intent in this section is to provide a basic description of the various types of transformation and profiling. In later chapters, we will dive into explicit examples of transformation and profiling.

Let's begin our discussion by exploring the transformation tasks involved in data wrangling. Table 3-1 describes the core types of data wrangling transformations that you might need to apply to your data.

Table 3-1. Core data wrangling types

Core transformation type	Description
Structuring	Actions that change the form or schema of your data
Enriching	Actions that add new values to your dataset
Cleansing	Actions that fix irregularities in your dataset

The core types of transformation are *structuring, enriching,* and *cleansing.* Structuring primarily involves moving record field values around, and in some cases summarizing those values. Structuring might be as simple as changing the order of fields within a record. More complex transformations that restructure each record independently include breaking record fields into smaller components or combining fields into complex structures. At the interrecord level, some structuring transformations remove subsets of records. Finally, the most complex interrecord structuring transformations involve aggregations and pivots of the data. Aggregations enable a shift in the granularity of the dataset (e.g., moving from individual customers to segments of customers, or from individual sales transactions to monthly or quarterly net revenue calculations). Pivoting involves shifting records into fields or shifting fields into records.

Whereas structuring transformations move or aggregate existing values from a single dataset, enriching transformations add fundamentally new values from multiple datasets. The quintessential structuring transformations are joins and unions. Joins combine datasets by linking records. Unions blend multiple datasets together by matching up records from two different datasets and concatenating them "horizontally" into a wider table that includes attributes from both sides of the match.

Beyond joins and unions, another common class of enriching transformations inserts metadata into your dataset. The inserted metadata might be dataset independent (e.g., the current time or the username of the person transforming the data) or specific to the dataset (e.g., filenames or locations of each record within the dataset). Yet another class of enriching transformations involves the computation of new data values from the existing data. In broad strokes, these kinds of transformations either derive generic metadata (e.g., time conversion or geo-based calculations like latitude-longitude coordinates from a street address or a sentiment score inferred from a customer support chat log) or custom metadata (e.g., mineral deposit volumes inferred from rock samples or health outcomes inferred from treatment records). Chapter 6 discusses specific examples of enriching transformations.

The third type of transformation cleans a dataset by fixing quality and consistency issues. Cleaning predominately involves manipulating individual field values within records. The most common variant fixes missing (or NULL) values. We explore specific examples of cleansing transformations in Chapter 7.

Switching gears, the core types of profiling are distinguishable by the unit of data they operate on: *individual values* or *sets of values.* Table 3-2 provides a description of the two types of core profiling.

Table 3-2. Two types of core profiling

Core profiling type	Description
Individual values profiling	Understanding the validity of individual record fields
Set-based profiling	Understanding the distribution of values for a given field across multiple records

Profiling on individual field values involves two kinds of constraints: syntactic and semantic. Syntactic constraints focus on formatting; for example, a date value should be in MM-DD-YYYY format. Semantic constraints are rooted in context or proprietary business logic; for example, your company is closed for business on New Year's Day so no transactions should exist on January 1 of any year. Ultimately, this type of profiling boils down to determining the validity of individual record field values, or, by extension, entire records.

Set-based profiling focuses on the shape and extent of the distribution of values found within a single record field, or in the range of relationships between multiple record fields. For example, you might expect retail sales to be higher in holiday months than in nonholiday months; thus, you could construct a set-based profile to confirm that sales are distributed across months as expected. We will explore specific examples of set-based profiling and individual profiling in Chapter 4.

So far, we've provided an overview of the basic types of transformations and profiling. Soon after doing the work, however, your focus will likely shift to include the second-order concern of doing the work well. In other words, in addition to putting a script of transformation logic together and profiling the steps of that script as you go to make sure they operate correctly on the initial data source(s), you want to optimize that script to run efficiently and robustly. Furthermore, over time, as new data mandates edits to the transformation script or you find more optimal ways to author portions of the transformation script, you'll likely want to track changes to the script and, possibly, manage multiple versions of it for legacy and capability purposes. Supporting these changes might require some additional, customized profiling information, tracking statistics across variations of the dataset.

Data Wrangling in the Workflow Framework

Data wrangling can be a major aspect of every action in your workflow framework (refer to Figure 2-2). In this section, we discuss each action in turn, describing how data wrangling commonly fits into the action.

Ingesting Data

As we discussed in Chapter 2, ingesting data into the raw data stage can involve some amount of data wrangling. Loading the data into the raw data stage location might require some nontrivial transformation of the data to ensure that it conforms to basic structural requirements (e.g., records and field values encoded in a particular formats). The extent of the constraints to load the data will vary by the kind of infrastructure of your raw data stage. Older data warehouses will likely require particular file formats and value encodings, whereas more modern infrastructures like MongoDB or HDFS will permit a wider variety of structures on the ingested data (involving less data wrangling at this stage).

In either event, the explicit goal when loading raw data is to perform the minimal amount of transformations to the data to make it available for metadata analysis and eventual refinement. The general objectives are "don't lose any data" and "fixing quality problems comes next." Satisfying these objectives will require limited structuring transformations and enough profiling to ensure that data was not lost or corrupted in the ingestion process.

Describing Data

Assessing the structure, granularity, accuracy, temporality, and scope of your data is a profiling-heavy activity. The range of profiling views of your data required to build a broad understanding of your data will also require an exploratory range of transformations. Most of the exploratory transformations will involve structuring: breaking out subcomponents of existing values to assess their quality and consistency, filtering the dataset down to subsets of records to assess scope and accuracy, aggregating and pivoting the data to triangulate values against other internal and external references, and so on.

Assessing Data Utility

Assessing the custom metadata of a dataset primarily involves enriching and cleaning transformations. In particular, if the dataset is a new installment to prior datasets, you will need to assess the ability to union the data. Additionally, you will likely want to join the new dataset to existing ones. Attempting this join will reveal issues with linking records between the datasets: perhaps too few links are found, or, equally problematic, there are too many duplicative links. In either case, by treating your existing data as a baseline standard to which the new dataset must adhere or align, you will likely spend a good amount of time cleaning and altering values in the new dataset to tune its overlap with existing data. As the new data is blended in with the old, set-based profiling will provide the basic feedback on the quality of the blend.

Designing and Building Refined Data

Building refined datasets that are broadly useful across a broad range of ad hoc analyses and deeper modeling and forecasting explorations requires the breadth of transformation and profiling types. Structuring the data to align with the granularity and scope of intended analyses will save time. For example, if most of your planned analyses are at a granularity (e.g., weekly total sales) that differs from the granularity of the raw data (e.g., individual sales transactions), it likely makes sense to apply aggregation or pivot transformations. In terms of enrichment, if many of your analyses involve multiple data sources, it makes sense to build blended datasets using enrichment transformations like joins and unions. Similarly, depending on how frequently your analyses require nontrivial derivations like smoothed time-series values or sentiment scores, it might make sense to build these into the refined datasets (as opposed to requiring each analysis to rebuild these enrichments as needed). Finally, cleaning the data is key to building broadly usable datasets. You need to flag inaccurate or inconsistent values as such (at a minimum) or replace them with more accurate/consistent versions. Likewise, many analyses will require missing data to be filled in with reasonable estimated values.

In terms of profiling, all types are required when building refined datasets. To ensure the quality and consistency of individual record values, profiling at this level should be aggressively applied across all record fields. Initially, for many fields, individual value profiling will enforce little more than syntactic constraints because semantic constraints on specific values will be unknown. As more and more ad hoc analyses are completed, and a better understanding of the value of the underlying data emerges, additional semantic profiling checks might be added. Similarly, as a deeper understanding of the dataset emerges through its use in many analyses, the richness of set-based profiling will increase. Initially, set-based constraints might enforce simple range or loose distribution checks. Over time, expected correlations between fields and trends in the changes of the distribution of field values (e.g., steady increasing of median sales prices) might be profiled and enforced.

It is difficult to overemphasize the criticality of building good refined datasets. From a governance perspective, these datasets will be the source of most of the insights that deliver value to you and your organization. The validity and consistency of these insights will depend on the quality of your refined datasets. In no small way, the trust an organization will have in the use of data to drive its decisions and operations will depend on the quality of these refined datasets.

Ad Hoc Reporting

Starting from refined datasets, reporting primarily involves structuring (or restructuring) input data. Perhaps you are exploring the impact of changing region boundaries and want to look at historical data to see if the new regions are more balanced in terms of traffic, sales, costs, and so on. Many of the necessary data transformations will involve pulling subcomponents out of record fields (for finer-grained analysis), filtering out nonrelevant records, and aggregating or pivoting metrics around subgroups of records. Additionally, depending on how your refined datasets are designed, ad hoc reporting might involve enrichment transformations, as well. Relative to the raw data ingested in the prior stage, much of the enrichment focused on joins and unions will already be done (though certainly not all). Beyond blending, data wrangling for ad hoc reporting might also require the derivation of new data values, perhaps involving the insertion of metadata like the current date, or calculating complex values like arbitrary percentiles or winsorized averages.

Exploratory Modeling and Forecasting

Like ad hoc reporting that builds from refined datasets, exploratory modeling and forecasting will employ a significant amount of structuring transformations. In addition to filtering, aggregating, and pivoting of records, it is common to pivot categorical record fields into separate indicator fields. This enables modeling techniques like regression. Furthermore, if your modeling analysis is focused on assessing the relative importance of various record fields, the fields might need to be normalized so that their corresponding model weights can be readily compared.

One of the benefits of exploring your data by using modeling and forecasting techniques is that these methods have the side effect of indicating when certain data points (record field values, or, by extension, entire records) are outliers; that is, they appear anomalous relative to the majority of the data. In some cases, these outlier data points might contain inaccurate values, instigating some additional data wrangling efforts focused on cleaning the data. In other cases, the outliers might represent valid data points that necessitate a change in how you understand the data or the processes that produce the data (e.g., most customers might spend only a few dollars per transaction, so if someone spends a few thousand dollars in a single transaction, what does it mean?).

Beyond building a single model or forecast, it is useful in these exploratory efforts to assess the robustness of the model or forecast—robustness relative to changes in the values of the input data (record field values), robustness relative to missing or deleted records, and so on. These kinds of robustness analyses involve transforming the original data (using structuring and cleaning type transformations) and then passing the modified data through the modeling and forecasting engine.

Building an Optimized Dataset

Similar to designing and building refined datasets, designing and building an optimized dataset involves the breadth of transformation types—structuring, enriching, and cleaning—and the breadth of profiling types—assessing both individual values and sets of values. The primary difference is in the balance of transformation to profiling. Whereas building refined datasets requires a fairly even mix of transformation and profiling, building optimized datasets requires significantly more profiling. It isn't enough to ensure that the transformation logic is correct for the specific dataset you just created. The transformation scripts that generate optimized datasets need to be automatically applied to regularly updating input data. Hence, you must use profiling to forecast the robustness of the transformation scripts to future variants of the data as well as to track the correctness of the optimized data each time the scripts are applied. Many of these profiling tasks will involve checking the distributions of values for various subsets of records—assessing both the range of values in the subsets as well as the shape of the distribution of values.

Regular Reporting and Building Data-Driven Products and Services

Analogous to the similarities and differences between building refined versus optimized data, regular reporting (relative to ad hoc reporting) and data-driven product and services (relative to exploratory modeling and forecasting) require similar transformations but more profiling. The driver for the additional profiling is the requirement that the same transformation scripts should work efficiently and robustly across evolving input data.

Figure 3-2 summarizes this section, highlighting the general amounts of different types of transformation and profiling across the actions in our workflow framework. In the figure, the bolded fonts represent the most frequently used transformation and profiling tasks for each action.

In the next few chapters, we discuss how you can apply the specific types of transformation and profiling actions to a sample data project.

Stage	Action	Transformations
Raw Data Stage	Ingesting Data	**Structuring** Enriching Cleaning **Individual Value Based Profiling** Set Based Profiling
Raw Data Stage	Generating Generic Metadata	**Structuring** Enriching Cleaning **Individual Value Based Profiling** **Set Based Profiling**
Refined Data Stage	Generating Proprietary Metadata	Structuring **Enriching** **Cleaning** Individual Value Based Profiling **Set Based Profiling**
Refined Data Stage	Designing and Building Refined Data	**Structuring** **Enriching** **Cleaning** **Individual Value Based Profiling** **Set Based Profiling**
Refined Data Stage	Ad-Hoc Reporting	**Structuring** Enriching Cleaning Individual Value Based Profiling Set Based Profiling
Refined Data Stage	Exploratory Modeling and Forecasting	**Structuring** Enriching Cleaning Individual Value Based Profiling Set Based Profiling
Production Data Stage	Designing and Building Optimized Data	Structuring Enriching Cleaning **Individual Value Based Profiling** **Set Based Profiling**
Production Data Stage	Regular Reporting	**Structuring** Enriching Cleaning **Individual Value Based Profiling** **Set Based Profiling**
Production Data Stage	Building Products and Services	**Structuring** Enriching Cleaning **Individual Value Based Profiling** **Set Based Profiling**

Figure 3-2. The relative amounts each type of transformation and profiling across the actions in our workflow framework

CHAPTER 4

Profiling

Overview of Profiling

We have decided to begin our discussion of data wrangling actions with profiling. This is the first action that people generally undertake when beginning each stage of a data project. Why? Because you need to understand the contents of your data before you can begin transforming or analyzing that data. Fundamentally, profiling guides data transformations.

When you're working on a data project, you often don't have time to look at every field of every record. Profiling is the activity that helps you know what is "in" your dataset, and allows you to validate that your transformations work as intended. Often, profiling is used to assess the quality of your data. Profiling is also a crucial aid for data transformation. You frequently need to be able to quickly determine if any records contain data that might cause problems during the transformation process. For example, if your downstream analysis expects each record in a price column to contain numbers, you don't want to have a record that includes letters or special characters.

Profiling can encompass two slightly different views:

- Examining individual values in your dataset
- Examining a summary view across multiple values in your dataset

Each of these views can often be consumed as textual information: a list of data values, a table of summary statistics, and so on. You can also build visualizations to capture profiling information about your data.

Ultimately, individual values profiling boils down to determining the validity of individual record field values. This type of profiling comes in two forms: syntactic checks and semantic checks. Syntactic constraints focus on formatting, and semantic constraints are rooted in context. Set-based profiling attempts to determine the validity of groups or distributions of values in a particular record field.

Beginning in this chapter, we will be working through an example data wrangling project using publicly available US campaign finance disclosure files. We will be wrangling candidate data and individual contribution data from the 2016 presidential election. The goal of this project is to see if there are any trends in the campaign contributions received by each of the two major candidates, Hillary Clinton and Donald Trump.

In each section, we reference the specific file that we are using so that you can follow the discussion in your data wrangling tool of choice. You can download all of the data files from *http://www.fec.gov/finance/disclosure/ftpdet.shtml*.

Individual Value Profiling: Syntactic Profiling

Syntax refers to constraints on the literal values that are valid in a field. The set of valid syntactic values might be quite small and best represented in a list; for example, Boolean values encoded as bits, {0, 1}, or birth sex encoded as English words, {male, female}. More often, syntactic constraints represent fairly large sets of possible values. For example, you might have a field that corresponds to the number of ATM cash withdrawals a customer has made in her lifetime (as a customer of your bank) in which the permissible values are integers ranging from 0 to 50000 (where the upper limit is determined by the permissible number of ATM cash withdrawals per day multiplied by the number of days your bank has been in business).

Profiling for syntactic constraints involves simply checking whether data values are in (or not in) the set of permissible values. At the specific example levels, a good choice for understanding syntactic type profiling is a random subset of values that satisfy the syntactic constraints along with a random subset of values that do not satisfy the syntactic constraints. From these examples, you often can make reasonable decisions as to whether a different syntactic type might be a better fit for a set of data.

Individual Value Profiling: Semantic Profiling

Semantic type constraints correspond to the meaning or interpretation of field values: values are valid if their interpretations satisfy the constraints. For example, suppose that a dataset has a field corresponding to the biological age, in years, of a customer. For some customers, suppose that we don't have a reported age, and the age field for the records corresponding to these customers contains the value -1. Although syntactically not a valid age in years, semantically this value can be interpreted as a Boolean indicator for whether or not the customer reported their age. Deriving a new field corresponding to reported age might be the variable we need to analyze how customers' willingness to report information predicts overall satisfaction.

As in the preceding example, the semantic interpretation of a field value might override syntactic constraints. As another example, consider a field encoding hometown/city. Suppose that some of the field values contain the strings "San Francisco, CA USA" and "Sn Francisco, CA, USA" and "San Francicso, CA, USA." While some of these are syntactically invalid, they are semantically clear and could be converted to their syntactically correct versions. Alternatively, suppose that the field values contain the string "Moscow." Should that be interpreted as Moscow, Russia, or Moscow, ID, USA? Perhaps there are other fields in the records that could disambiguate between these cases, but on their own, these values would be semantically invalid.

In other cases, the semantic interpretation of a field might involve a simple conversion from one categorical space to another; for example, age in years to basic life stages (e.g., teenager, young adult, senior), or time of day to basic day stage (e.g., morning, afternoon, evening, and night).

Profiling semantic type constraints often requires deriving a new record field that explicitly encodes the semantic interpretation of a source field. This explicit encoding can then be syntactically typed, where validity can be determined by testing whether the value is in the valid set. Thus, as with syntactic types, the most appropriate summary statistics correspond to the percentage of field values that are valid, invalid, and empty/null.

The common, and more straightforward, case uses deterministic rules for converting a source field to its interpreted value. For example, you might define a rule stating that anyone between the ages of 13 and 19, inclusively, is a teenager. The more difficult case involves nondeterministic, or probabilistic, mappings from source values to interpretations. For example, based on summary demographics that we know about our dataset, we might interpret "Moscow" as 80 percent likely to represent Moscow, Russia, and 20 percent likely to represent Moscow, ID, USA.

Set-Based Profiling

Set-based profiling focuses on the shape and extent of the distribution of values found within a single record field or in the range of relationships between multiple record fields. For numeric fields, distributional profiling often builds from a simple histogram of the set of values and might involve comparing that histogram against a known distribution like a Poisson or Gaussian probability distribution.

In addition to looking at the overall distribution, it helps to look at various summary statistics like minimum, maximum, mean, and sum. These values can provide you with an understanding of the distribution of values across your dataset and help you to immediately identify any problematic distributions or outliers.

For categorical record fields, a couple of different distributional profiles are useful. The first counts occurrences of unique values. Another useful profiling chart for categorical variables clusters the raw values, via a mechanism similar to standardization, and then counts the number of values associated with each cluster.

For more specific types of record fields, there are more specific profiling charts. For example, for geospatial data like zip codes or latitude-longitude coordinates, plotting the data on a map is appropriate.

For date-time data, it is useful to see the values plotted on a variety of scales. You could examine the distribution of date-time values across the 24-hour day, across the 7 days of the week, or across the 12 months of the year.

In addition to distributional profiling of the set of values from a single record field, cross-distributional profiling of values from multiple record fields is often useful in assessing the quality of the overall dataset. Simple scatterplots are commonly utilized.

Profiling Individual Values in the Candidate Master File

 To follow along yourself, download the "Candidate Master File" from election year 2015-2016 (*http://www.fec.gov/finance/disclo sure/ftpdet.shtml#a2015_2016*).

You also might need to reference the data dictionary for this file to understand the permissible values in each field. You can find the data dictionary online (*http://www.fec.gov/finance/disclosure/meta data/DataDictionaryCandidateMaster.shtml*).

As noted in the introduction to this chapter, we will be attempting to wrangle US campaign finance data to build a refined dataset that will help answer the question: "How were individual campaign contributions distributed between the two 2016 major party US presidential candidates?"

Profiling the available datasets is the first step toward constructing a refined dataset that maps individual campaign contributions to each 2016 presidential candidate. The profiling step is particularly important for this project because the data is largely unknown. Data projects that use publicly available data often require a fairly lengthy profiling process to familiarize yourself with the contents of the dataset.

To help the public understand the contents of these datasets, the US Federal Election Commission (FEC) has provided a data dictionary to supplement the raw candidate and contributions files. We can use the FEC's data dictionary to inform our profiling of individual values. If you examine the data dictionary, you can see the constraints that define permissible values for each column in the dataset.

Data dictionaries are frequently generated as the output of the metadata-producing actions during the raw data stage. Data dictionaries define the permissible valid values in each column and can also explain the contents of the dataset.

We're going to begin by looking at the "Candidate Master File" for election year 2015–2016. In case you're not familiar with US election rules, this file contains information about all of the candidates who filed the appropriate form to be included in an upcoming election. It also includes candidates with active election committees, even if the election that they have registered for is not the current election. As you examine this file, you might also notice some creative candidate names; even though these candidates submitted paperwork to be included in an election, they might not have actually appeared on individual state ballots.

We want to use profiling to answer two specific questions:

- Are there any values in columns that are syntactically invalid given the scope of our project?
- Is the range of values in each column valid given the scope of our project?

Syntactic Profiling in the Candidate Master File

Let's begin by profiling the two US state columns. We want to assess whether the values contained in those columns are syntactically valid. In the FEC's data dictionary, column 5 represents the state of the office for which the candidate is running, and column 14 represents the state of the candidate's mailing address.

Focusing on column 5, we can begin by collecting a list of all the unique values in the column. This is a common first step when performing syntactic profiling. This operation produces a list of 57 unique values. We know that there are 50 US states that have voting representatives in the US Congress, and 5 US territories and the District of Columbia with nonvoting representatives in the US Congress. Additionally, if we look at the data dictionary, we can see that column 5 can contain a nonstate value, "US," in records that represent candidates who are running for president. So, at first glance, it seems reasonable that there would be 57 possible syntactically valid locations in column 5.

We can dig a little deeper and examine each value in column 5 individually to see if it matches one of the known 50 state abbreviations, 6 territory abbreviations, or "US." We performed this check by using a lookup to a reference dataset of all 57 valid values. In fields that matched one of the 57 valid values, we inserted a "1," and in fields that did not match one of the 57 valid values, we inserted a "0." Ultimately, all the values in this column are syntactically valid; in our Boolean indicator column, 100 percent of the records contained "1."

Because column 14 also contains state abbreviations, we can perform a similar set of profiling checks on this column. Again, a count of all the distinct values in this column reveals that there are 57 possible values. However, because this column represents a mailing address, there are only 56 possible valid values: 50 US state abbreviations, "DC" for the District of Columbia, and 5 US territory abbreviations. At first glance, we can assume that at least some of the records contain syntactically invalid entries in column 14. In addition to 57 distinct values, this column also contains missing values. We can consider missing values syntactically invalid because the data dictionary does not indicate that missing values should appear in this column.

We'll use the same procedure that we applied when profiling column 5 to see which of the individual values in column 14 are syntactically valid. Performing a lookup to a reference table and generating a Boolean indicator column shows that there is a single record that contains an erroneous state: "ZZ."

You can perform a similar set of syntactic checks on the other columns in the Candidate Master File. We recommend generating a series of Boolean indicator columns to show whether the values in each record are permissible given the constraints defined in the FEC's data dictionary.

Set-Based Profiling in the Candidate Master File

Let's profile the distribution of values in column 4 of the Candidate Master File. According to the data dictionary, this column represents the year of the election for which each candidate registered. Since this dataset can include candidates for any election with active campaign committees, we would expect to see the years distributed so that there are relatively few records for elections prior to 2016, a large number of records for the 2016 election year, and possibly a small number of records that represent future elections (perhaps 2018 or 2020).

After you've generated a summary view that counts the number of records that occur in each year, you should see a very wide range of values in column 4. The earliest recorded date is 1990; the date farthest in the future is 2064.

At this point, we would recommend stepping back to determine the utility of records in this column. If you remember our discussion in Chapter 2, assessing the utility of your data involves generating custom metadata, or metadata specific to your use case. That means that we should assess the distribution of the values in column 4 in the context of our specific project to see how many of these records are relevant to our analysis. The goal of this project is to see if there are any trends in the campaign contributions received by each of the two major candidates in the 2016 presidential election, Hillary Clinton and Donald Trump. Since we're interested in only the 2016 presidential election. records that represent candidates registered for the elections in 1990 or 2064 are ultimately irrelevant to our task. We can insert additional metadata into our dataset at this stage, perhaps flagging records that contain a value other than "2016" in column 4 as invalid.

You can profile the values in column 6 of the Candidate Master File in a similar way. This column represents each candidate's desired office. Based on our project's constraints, attempt to make an assessment about the utility of each category of values in this column.

CHAPTER 5

Transformation: Structuring

Overview of Structuring

You might remember our discussion of *structure* as a metadata element from Chapter 2. Structuring as a transformation action involves changing your dataset's structure or granularity. In other words, structuring consists of any actions that change the form or schema of your data.

At a high level, there are two sets of structuring actions that you might need to apply to your datasets. The first group of structuring transformations involves manipulating individual records and fields. We call this *intrarecord structuring*. Intrarecord structuring transformations roughly fall into three buckets:

- Reordering record fields (moving columns)
- Creating new record fields through extracting values
- Combining multiple record fields into a single record field

The second group of structuring transformations involves operating on multiple records and fields at once. We call this *interrecord structuring*. These types of transformations fit roughly into two types:

- Filtering datasets by removing sets of records
- Shifting the granularity of the dataset and the fields associated with records through aggregations and pivots

We will discuss each set of structuring transformations in turn so you can understand when you might want to apply these operations to your datasets.

Intrarecord Structuring: Extracting Values

Extraction involves creating a new record field from an existing record field. Frequently, this involves identifying a substring in an existing column and placing that substring into a new column.

Positional Extraction

The simplest form of substring extraction works by specifying the starting position and ending position that correspond to the substring that you want to extract from a set of record fields. This is called positional extraction.

When you're working with data, extracting substrings based on a consistent position is common when dealing with date-time fields or fixed-width fields. Both of these field types have known elements located at specific positions, so there is generally little to no inconsistency in the structure of the field.

 To follow along yourself, download the "Contributions by Individuals" file from election year 2015-2016. You can find the file at *http://www.fec.gov/finance/disclosure/ftpdet.shtml#a2015_2016*.

You also might need to reference the data dictionary for this file to understand the permissible values in each field. You can find the data dictionary at *http://www.fec.gov/finance/disclosure/metadata/ DataDictionaryContributionsbyIndividuals.shtml*.

Let's look at an example of a field for which positional extraction might be a valuable structuring technique. In the Individual Contributions dataset, column 14 contains the transaction date for each campaign contribution. In this example, we want to extract the day of the month for each individual contribution into a new column. This will allow us to create a field that we can use to determine if individual campaign contributions were more frequent in certain times of the month.

The following table shows four example record fields from column 14:

column14
03102015
03302015
03302015
03022015

If you look at the source data, you can see that the discrete elements of the field always align: the month occurs first, the day of the month follows after exactly two characters, and the year follows after exactly four characters. Each individual record field is homogeneously structured. This means that positional extraction will allow us to easily identify the block of text that represents the month of the contribution.

To identify the starting and ending positions of the substring that represent the day of the month, we can map each individual character in the source record field to a position. You can see how this will work here:

```
0        3  1  0  2  0  1  5
Position 1  2  3  4  5  6  7  8
```

By counting the individual characters, including spaces, we can see that the day of the month starts at position 3 and ends at position 4.

In a sentence, we could describe our desired transformation by saying, "From source column 14, extract the characters located from position 3 to position 4." If we were to perform this transformation on the dataset, we would produce the following output:

column14	Day of Month
03102015	10
03302015	30
03302015	30
03022015	02

A more complex version of positional substring extraction can pluck a substring from a record field when the starting and ending positions of the substring differ from record to record. Address fields are an excellent example of record fields for which complex positional extraction can be utilized effectively.

To perform this type of positional extraction, you will want to use functions that can search for a particular sequence of characters within the original record field. These functions return either the start position of the searched-for sequence or the length of the sequence. You can then pass the values returned by these functions through to one of the basic positional extraction functions. This will produce a complex nested function.

Looking again at the Individual Contributions dataset, you can see that column 8 contains the name of the person or organization who made each campaign contribution. For records that represent individual people, commas separate the person's first name and last name. A sample of data from column 8 is shown here:

```
ARNOLD, ROBERT
BICKLE, DON
ROSSMAN, RICHARD
LLEWELLYN, CHARLES
```

You can see that the last name element has an inconsistent length in each record. In the first record, the last name is 6 characters long (ARNOLD), whereas in the third record, the last name is 7 characters long (ROSSMAN). Simple positional extraction would not work in this case because the ending positions of the last name differ from record to record. However, because the first name and last names in each record are all separated by a common delimiter—the comma—we can use complex positional extraction functions to identify the position of the comma and then extract the appropriate substring.

Pattern Extraction

Pattern-based extraction is another common method that you can use to extract substrings into a new column. This method uses rules to describe the sequence of characters that you want to extract. To explain what we mean, let's look at another sample of data from the Individual Contributions dataset. According to the FEC's data dictionary, column 20 contains free text that describes each contribution. A sample of data from this column is below:

column20

```
P/R DEDUCTION ($296.67 MONTHLY)
P/R DEDUCTION ($326.67 MONTHLY)
* EARMARKED CONTRIBUTION: SEE BELOW
P/R DEDUCTION ($1000.00 MONTHLY)
```

In this case, we want to extract the monthly contribution amount into a new column. It's fairly easy to describe the desired transformation in a sentence: "From column 20, extract the first sequence of digits, followed by a period, followed by another sequence of digits." In this sentence, the pattern that defines the street name reads, "first sequence of digits, followed by a period, followed by another sequence of digits." You can often use regular expressions to represent patterns in code. Regular expressions are also supported by most data wrangling software products.

As you can see, pattern-based extraction can be a useful method to identify substrings that conform to the same generic pattern but are not identical.

Complex Structure Extraction

Sometimes, when you are wrangling data, you might need to extract elements from within complex hierarchical structures. We commonly see this type of complex structure extraction required when wrangling JSON files or other semistructured formats. (You can refer to our discussion of *structure* in Chapter 2 for a refresher on the differences between structured and semistructured data.) JSON-formatted data often originates from automated systems; if you are working with machine-generated data, it's likely that your data contains JSON structures.

Users who are wrangling JSON data generally encounter two types of complex structures: maps and arrays. These structures are common in semistructured data because they allow datasets to include a variable number of records and fields. They are described here:

JSON array
> A JSON array represents an ordered sequence of values. JSON arrays are enclosed in square brackets. Elements in arrays are separated by commas and enclosed in double quotes.

> Example array: ["Sally","Bob","Alon","Georgia"]

JSON map
> A JSON map contains a set of key-value pairs. In each key-value pair, the key represents the name of a property and the value represents the value that describes that property. JSON maps are enclosed in curly brackets. Key-value pairs are separated by commas. Keys and values are both enclosed in double quotes.

> Example map: {"product":"Trifacta Wrangler","price":"free","category":"wrangling tool"}

In a given dataset, an array in one record might be a different length from an array in another record. You might see this in a dataset containing customer orders, where each record represents a unique customer's shopping cart. In the first record an array of orders might include two elements, whereas in the next record, an array of orders might include three elements.

Similarly, maps also support variability across records. Looking at the shopping cart example, each cart might contain a variety of possible properties—say, "gift_wrapped", "shipping_address", "billing_address", "billing_name", and "shipping_name". Ideally, every record will contain all of the possible properties. However, it's more likely that some shopping carts only contain a subset of possible properties. Representing the properties and their associated values in a JSON map allows us to avoid creating a very sparsely populated table.

Of course, JSON format, although ideal for storing data efficiently, is often not structured ideally for use in analytics tools. These tools commonly expect tabular data as input. Consequently, when working with a JSON array or map, you might need to pluck a single element into a new column, or fold the multiple elements contained in an array down into multiple records. This will allow you to convert JSON-formatted data into the rectangular structure needed for downstream analytics.

When to Use Each Substring Extraction Technique

Simple positional extraction
> Use this when each record field conforms to a known structure, and each unique substring in the field always begins and ends at the same position. Fixed-width fields and date-time fields are good candidates for positional extraction.

Complex positional extraction
> Use this when each record field contains multiple substrings that are separated by the same delimiter string. Substrings do not necessarily need to be the same length or conform to the same pattern. Address fields and full name fields are good candidates for complex positional extraction.

Pattern-based extraction
> Use this when you want to extract a substring that can be defined by a generic pattern.

Complex structure extraction
> Use this when your record fields contain JSON maps or arrays, and you want to extract individual elements from those structures. Many datasets that are machine-generated contain complex JSON structures.

Intrarecord Structuring: Combining Multiple Record Fields

Combining multiple fields is essentially the reverse of extraction. When you are wrangling data, you might need to create a single field that merges values from multiple related fields.

As an example, let's return to the Individual Contributions dataset. This dataset contains two related columns: column 9 (city) and column 10 (state).

You can see that the city column adds additional detail to the state column. We want to combine the data from these two columns into a single column, and then separate the city and state with a comma. Combining the data from the these two fields can be useful if your downstream analysis wants to consider this data as part of a single record field.

Our desired output will look like the following column:

City State
MCPHERSON, KS
FREDERICK, MD
BROKEN BOW, NE
HAWTHORNE, CA

Interrecord Structuring: Filtering Records and Fields

Filtering involves removing records or fields from a dataset. Although filtering is often utilized in cleaning transformations designed to address dataset quality (which we discuss further in Chapter 7), you also can use it to alter the granularity of a dataset by changing the types of records and fields represented in a dataset.

For example, the Individual Contributions dataset contains a column that represents the type of entity that made each donation. Based on the FEC data dictionary, this field contains eight distinct values: CAN, CCM, COM, IND, ORG, PAC, and PTY. Based on this column, we could say that the granularity of the dataset is fairly coarse. After all, records can belong to one of eight distinct groups.

Let's assume that we are interested in analyzing only campaign contributions that originated from individuals (represented in the entity column by "IND"). We will need to filter our dataset so that it includes records that contain only the value "IND" in column 7. Performing this operation will produce a dataset with a finer granularity because each record will now belong to only a single category of values from the entity type column. This type of filtering is called *record-based filtering.*

Another type of filtering that is commonly used as a structuring operation is *field-based filtering.* This type of filtering affects the number of fields, or columns, in your dataset.

Interrecord Structuring: Aggregations and Pivots

Aggregations and pivots are structuring operations that enable a shift in the granularity of a dataset. For example, you might start with a dataset of sales transactions and want total sales amounts by week or by store or by region—a fairly straightforward aggregation involving the summation of record fields. A more complex pivot might involve extracting the items purchased out of the transaction records and building a dataset in which each record corresponds to an item.

For example, consider a dataset composed of individual sales transactions, where each transaction record contains a field listing the products that were purchased. You can pivot this dataset such that each product becomes a record with fields describing the product and an aggregated count field indicating the number of transactions involving this product. Alternatively, you could pivot the same dataset to count the number of transactions per product where the product was purchased alone, with one additional product, with two additional products, and so on.

To coordinate our discussion in this section, we can organize aggregations and pivots into three progressively more complex groups. These groups can be characterized by the relationship between records and record fields between the input dataset (prior to applying the transformation) and the output dataset.

Simple Aggregations

In the first group, simple aggregations, each input record maps to one and only one output record, whereas each output record combines one or more input records. For simple aggregations, the output record fields are simple aggregations (sum, mean, min, list concatenation, etc.) of the input record fields.

We can perform a basic aggregation on the Individual Contributions dataset. Perhaps we want to manipulate the granularity of the dataset so that each row summarizes the campaign contributions made to each campaign committee. We are interested in creating three new columns:

- One column that contains the average contribution made to each campaign committee
- One column that contains the sum of contributions made to each campaign committee
- One column that counts the number of contributions made to each campaign committee

In this example, we will be performing this basic aggregation on the following limited sample of data from the Individual Contributions dataset:

Column 1	Column 15
C00004606	750
C00004606	1000
C00452383	225
C00452383	50

Remember, based on the FEC's data dictionary, column 1 contains the campaign committee and column 15 contains each contribution amount. After aggregating the values in column 15 by campaign committee, we end up with the following output:

Column 1	Sum of Column 15	Mean of Column 15	Count of Column 15
C00004606	1750	875	2
C00452383	275	137.50	2

Column-to-Row Pivots

In the second group, column-to-row pivots, each input record maps to multiple output records, and each output record maps to one and only one input record. Input record field values become the defining characteristics of the output records. In other words, the output records contain a subset of the input record fields.

This type of column-to-row pivot is commonly referred to as "unpivoting" or "denormalizing" data. It is particularly useful when your source data contains multiple columns that represent the same type of data. For example, you may have a transactions file that contains the total sales numbers per region, per year. The data could be formatted as shown in the following table:

Region	2015	2016
East	2300	2453
West	9866	8822
Midwest	2541	2575

Note that in this example, the sales figures for each year are contained in a different column. We want to restructure this dataset so that a single row contains the sales for a single unique combination of region and year. The result of this column-to-row pivot will look like the following table:

Region	Year	Sales
East	2015	2300
East	2016	2453
West	2015	9866
West	2016	8822
Midwest	2015	2541
Midwest	2015	2575

Generally, column-to-row pivots are used when you want to create a dataset that allows you to more easily summarize values. Compared to the original sales dataset in our example, the resulting dataset is structured to facilitate calculations across years and regions.

Row-to-Column Pivots

In the final group, output records sourced from multiple input records and input records might support multiple output records. Output record fields might involve simple aggregations (e.g., sum or max) or involve more complex expansions based on the field values. This type of pivot is called a row-to-column pivot.

As an example, let's return to the Individual Contributions dataset. We want to create a refined dataset that shows the sum of contributions made to each campaign committee, broken out by contribution type. In this case, we want to create one new column for each contribution type.

A subset of the Individual Contributions dataset contains the following data:

Column 1	Column 7	Column 15
C00004606	IND	750
C00004606	IND	1000
C00492116	PAC	45000
C00492116	PAC	15000
C00492116	IND	250
C00452383	750	50

Based on the FEC's data dictionary, column 1 represents the campaign committee, column 7 represents the contribution type (individual, political action committee, corporate, and so on), and column 15 represents the contribution amount. After performing a row-to-column pivot, our subset of the Individual Contributions dataset will look like the following table:

Column 1	Sum of IND Contributions	Sum of PAC Contributions
C00004606	1750	0
C00492116	60000	1000

As a result of the row-to-column pivot, we have created a new column for each of the unique values in source column 7. Each row in this new dataset summarizes contributions made to a single campaign committee.

CHAPTER 6

Transformation: Enriching

Enriching transformation actions result in the net addition of information to your dataset. When enriching your dataset, you insert additional records or fields from other related datasets, or you use formulas to calculate new fields.

You might wonder how enriching transformations differ from structuring transformations (discussed in Chapter 5). Although both types of transformations can involve creating new fields or records, structuring transformations create new fields or records based on data already present in the dataset. Enriching transformations, in contrast, create new fields or records using new data—information that was not previously present in the dataset in any form.

There are three primary types of enriching transformations:

- Unions
- Joins
- Deriving new fields

We discuss each type of enriching transformation in this chapter.

Unions

Unions append additional records to an existing dataset. In other words, when you perform a union, you are taking two related datasets and stacking them vertically to create a single dataset.

Why might you want to perform a union? Let's imagine that you work for an organization that receives monthly orders from your clients. At the end of each quarter, you need to produce a summary analysis that records the total number of orders placed by each client over the previous three months. Because each month's orders are contained in a separate dataset, you need to combine them into a single dataset so that you can perform your analysis.

The simple case is when records from both the old and new datasets have matching structures (i.e., matching layout of fields for each record). In this case, a union amounts to little more than a concatenation of datasets.

The more complex—and common—case is when the records from the current dataset have a different structure from the records in the new dataset. Generally, the difference is minor: most of the fields are in the same layout and there are only a few fields that differ between the old and new data records. In this case, some simple logic can dictate whether the unmatched fields should be coerced into the same field or kept in separate fields with null or empty values inserted into the records from the other dataset.

Joins

Joins are the most common enrichment action. The most common form of a join involves linking records from one dataset to records from the other dataset via exact matches of a single field in each of the dataset records. The field used in the match is often referred to as the *key field* in both datasets. For example, consider a dataset containing customer profile information with a key field corresponding to a customer_id. In a second dataset, suppose that we have customer transaction information along with a customer_id key field. By joining records on customer_id matches, we can now analyze how transaction activity relates to profile information.

In most tools, the link between two datasets is based on exactly matching one or more fields between the records. Some tools additionally support fuzzy matching of fields. This allows records with misspellings or other minor differences to be linked.

There are four types of joining/blending logic. They differ in how they handle the variety of situations that arise when matching records for the two datasets. Obviously, when records match from the joining datasets, we simply append record fields (or a subset of records fields). It is common to think about the case in which every record from each dataset has one and only one matching record in the other dataset. What happens when a record from one dataset matches multiple records from the other dataset? Or when records from one dataset have no matching records from the other dataset?

Let's take a closer look at the four types of joins:

Inner
> Inner joins only produce a record when there are matching records from each dataset being blended. Note that if there are duplicate keys, the output records are similarly duplicated.

Left outer

> This type of join retains all records from the left (or initial) dataset, even if there is no matching record in the right (or incoming) dataset.

Right outer

> These joins retain all records from the right (or incoming) dataset, even if there is no matching record in the left (or initial) dataset.

Full outer

> Full outer joins retain all records from both datasets, even if they have a corresponding match.

Inserting Metadata

An important form of enrichment involves adding metadata into the dataset. Common metadata to add include the filenames of the source data, byte offsets and/or record numbers, current date and/or time, creation/update/access timestamps, and record and/or record field lineage.

Derivation of Values

A final kind of enrichment involves the derivation of new values. There are two basic kinds of derivation: *generic* and *proprietary*.

Generic

Generic derivations apply to many datasets. For example, most datasets need to explicitly address time—either by encoding temporal information within records or across records as metadata. A common data wrangling action involves deriving additional date-time information; for example, day of the week or season. Or, when data is collected across multiple time zones, it can be useful to derive both a local and global (e.g., UTC) timestamp for each event. These derived values can be crucial to an analysis that deals with weekly or yearly cycles, experiential time, and absolute sequencing of events (e.g., for anomaly detection).

Another common domain for many analyses involves geography or spatial encodings. Sometimes, it is as simple as abstracting an address into a ZIP code or city. Or, utilizing many available services, you could also convert an address to latitude and longitude coordinates. Slightly more involved derivations might convert an address to a more customized region relevant for marketing or sales activity. These regions might be drawn, for example, by calculating the shortest driving distance between a customer and all available nearby stores. None of the three data wrangling tools support geographic and spatial encodings; you would need to create custom code to produce these features.

A third generic domain for derivation functionality concerns text. Generally referred to as Natural Language Processing, or NLP, functionality in this domain could take raw text and derive some kind of overall sentiment analysis; extract out references to people, events, or things; or identify topics that capture the text's semantics. You might also want to translate text from one language to another. When it comes to NLP, a key concern is what defines your baseline, or reference language; for example, are you working with only English text, or are you working with terminology specific to your industry? Choosing an appropriate reference language will determine how you treat jargon or slang, for example.

Finally, a fourth generic domain for derivation functionality concerns basic numeric calculations. Some of these calculations are simple; for example, summing the list price and taxes of a transaction to produce a final price. Others are slightly more sophisticated, utilizing aggregation or sequential processing mechanisms. An example would be z-normalizing a numeric field by subtracting the mean value of the field and then dividing by the standard deviation of the field.

Some of the generic derivation functionality is domain specific. For example, you might have different region definitions if you are a mining company versus a consumer packaged goods retailer. Similarly, if you are performing NLP derivations, you can target jargon or terminology specific to your industry (like silicon manufacturing, pharmaceuticals, fashion, etc.).

Legal requirements around report compliance and privacy drive additional domain-specific derivations. For example, in the healthcare domain, laws pertaining to patient privacy and confidentiality require certain record fields to be removed and replaced with derived ones. Also, given a history of treatments, one might derive record fields indicating whether certain future treatments are viable or not. Similarly, in finance, there are domain-specific derivations related to the time series of transaction data that drives that industry. Specific models, such as Black-Scholes, can be applied to historical time series of trading data to predict future value. These predictions can be analyzed along with other predictions to identify the optimal model to use or to build a composite prediction.

Proprietary

Further along the generic-to-specific spectrum are proprietary derivations. In this case, individual organizations might have custom models they use to make predictions, for example. Or, they might have highly customized derivation calculations that best capture, for example, the health of their customer base or the likelihood of a customer leaving or upgrading.

In many big-data data wrangling systems, proprietary functionality is often encoded as User-Defined Functions (UDFs). Moreover, these UDFs are often supported in a variety of computational languages: R, Python, VisualBasic, Java, and so on. Whereas many UDFs contain nontrivial calculations, others make calls to services (like Google's geocoding API, which converts addresses to latitude and longitude coordinates).

CHAPTER 7

Using Transformation to Clean Data

The third type of data transformation cleans a dataset to fix quality and consistency issues. Cleaning predominately involves manipulating individual field values within records. The most common variants of cleaning involve addressing missing (or NULL) values and addressing invalid values.

Addressing Missing/NULL Values

There are two basic approaches to addressing missing/null values. On the one hand, you can filter out records with missing or NULL fields. On the other hand, you can replace missing or NULL values. Often referred to as data imputation, filling in missing or NULL values might utilize many different strategies. In some cases, the best approach involves inserting the average or median value. In other cases, it is better to generate values from similar records; for example, similar customers or similar transactions. Alternatively, if your data has strong ordering (because it is a time-series dataset, for example), you might be able to fill in missing values by using the last valid value.

Addressing Invalid Values

Extending beyond missing values, another key set of cleaning transformations deals with invalid values—invalid because they are inconsistent with other fields (e.g., a customer age compared with their data of birth), ambiguous (e.g., two-digit years or abbreviations like "CT"—is that Connecticut or Court?), or improperly encoded. In some cases, the correct or consistent value for the field can be calculated and used to overwrite the original value in the dataset. In other cases, it might make sense for you to simply mark values as invalid. You can then conduct two parallel analyses, one that includes the invalid values and one that excludes them, providing insight into the impact that invalid data is having on your insights.

A more complex variety of fixing invalid values involves data standardization. Furthermore, suppose that every customer represented in that dataset is known to reside in the United States. A reasonable validity check on the current-state-of-residence field is that it should fall into one of the known US states. Suppose, however, that there are misspellings: "Californa," "Westvirginia," and "Dakota." Standardizing these field values to a fixed library of valid values is a good way to improve dataset quality. There are a number of ways to perform this kind of standardization. The most common method involves editing distance around misspelling; that is, strings that are similar, like "Californa" and "California," should be treated as the same entity and converted to the same spelling.

More specific standardization techniques rely on domain knowledge. For example, is "Dakota" supposed to be "North Dakota" or "South Dakota"? If we have a ZIP code in another field of the record, perhaps we can use a mapping of ZIP codes to states to make this determination. A slightly less reliable mapping, now that cell phone numbers can be transferred across carriers, could use the area code on a customer phone number field.

CHAPTER 8

Roles and Responsibilities

There are a number of common job titles and roles for people who work with data. In our experience, the most common are *data engineer*, *data architect*, *data scientist*, and *analyst*. In this chapter, we provide a general overview of each of these roles, which actions in the workflow they tend to be responsible for, and some best practices for managing the hand-off between these roles and for driving the long-term success of your data practices.

Skills and Responsibilities

We'll provide a basic overview of the four common job roles that we encounter when working with data. Of course, in smaller organizations or in personal projects, a single person can end up developing and applying all of the skills and responsibilities that we'll discuss. However, it's more common to split them into separate job roles.

Our discussion is oriented on two axes (see Figure 8-1). The first axis is focused on the primary kind of output produced by someone in the role. The second axis is focused on the skills and methods utilized to produce that output. We'll discuss each role in turn.

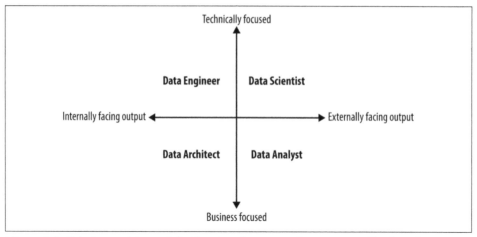

Figure 8-1. Relative positions of the four key data wrangling user profiles based on output (internal or external) and skills (technically focused or business focused)

Data Engineer

Data engineers are responsible for the creation and upkeep of the systems that store, process, and move data, as shown in Figure 8-2. In addition to instantiating and maintaining these systems, many data engineers focus on the efficiency and extensibility of these systems, ensuring that they have sufficient capacity for existing and exploratory or future workloads.

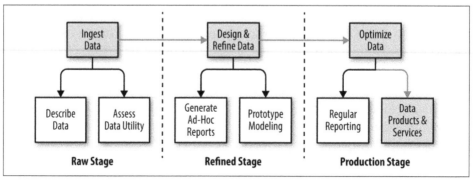

Figure 8-2. Blue highlights identify the primary actions for a data engineer in our data wrangling workflow framework

The ability to create and maintain these systems requires some background in system administration. More importantly, as these systems are primarily designed to work with data, data engineers require fairly deep background in common data processing algorithms and implementation of these algorithms across various systems and tools.

Data Architect

As Figure 8-3 illustrates, data architects are responsible for the data in the "refined" and "production" stage locations (occasionally they are responsible for the "raw" data, as well). Their objective is to make this data accessible and broadly usable for a wide range of analyses. In addition to staging the data itself, data architects often create catalogs for this data to improve its discoverability and usability. Further optimizations to the data and the catalog involve the creation of naming conventions and standard documentation practices, and then applying and enforcing these practices.

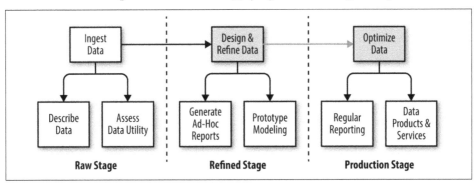

Figure 8-3. Red highlights identify the primary actions for a data architect in our data wrangling workflow framework

In terms of skills and methods, data architects often work through a user-requirements-gathering process that moves from documented data needs and requests, to an abstract model of where to source the data and how to organize it for broad usability, to the concrete staging of the data by designing data schemas and alignment conventions between the schemas. Designing the structure, ensuring the quality, and cataloging the relationships between these dataset builds on fluency in the data access and manipulation languages of a broad set of data tools and warehouses (e.g., variants of SQL, modern tools like Sqoop and Kafka, and more analytics systems like those built by SAP and IBM). Additionally, data architects often employ standard database conventions like First, Second, and Third Normal Forms.

Data Scientist

Data scientists are responsible for finding and verifying deep or complex sets of insights (Figure 8-4). These insights might derive from existing data using advanced statistical analyses or from the application of machine learning algorithms. In other cases, data scientists are responsible for conducting experiments, like modern A/B tests. In some organizations, data scientists are also responsible for "productionalizing" these insights.

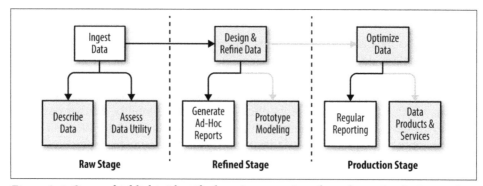

Figure 8-4. Orange highlights identify the primary actions for a data scientist in our data wrangling workflow framework

There are two primary types of data scientists: those who are more statistics focused and those who are more engineering focused. Both are generally tasked with finding deep and complex insights. Where they differ is in the last couple of just-listed responsibilities. In particular, more statistics-focused data scientists will often focus on A/B testing, whereas those who are more engineering-focused will often concentrate on prototyping and building data-driven services and products.

The ability to identify and validate deep or complex insights requires some familiarity with the mathematics and statistics algorithms that can reveal and test these insights. Additionally, data scientists require the skills to operate the tools that can apply these algorithms, such as R, SAS, Python, SPSS, and so on. For statistics-focused data scientists, background in the theory and practice of setting up and analyzing experiments is a common skill. For engineering-focused data scientists, skills around software engineering are required, not just familiarity with a variety of programming languages, but also with best practices around building complex applications.

Analyst

Analysts are responsible for finding and delivering actionable insight from data, as depicted in Figure 8-5. Whereas data scientists are often tasked with exploratory, open-ended analyses, analysts are responsible for providing a business or organization with critical information. In some cases, these take the form of top-line metrics, KPIs, to drive or orient the organization. Sometimes these metrics are delivered in reports (both regular and ad hoc); other times they are delivered as general talking points to help justify a decision or course of action. The line between an analyst and a data scientist can be blurry. In many situations, analysts will pursue deeper analysis. For example, in addition to identifying correlated indicators of KPI trends, they might perform causal analysis on these indicators to better understand the underlying dynamics of the system.

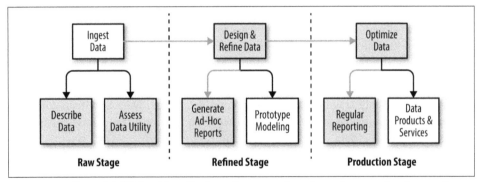

Figure 8-5. Green highlights identify the primary actions for a data analyst in our data wrangling workflow framework

Along those lines, in addition to good mathematics and statistics backgrounds, analysts are often steeped in domain expertise. For most organizations, this amounts to a deep understanding of the business and marketplace. More generically, analysts are strong systems thinkers; they are able to connect insights that might be corelevant and then propose ways to measure the extent of their relationships.

Roles Across the Data Workflow Framework

The workflow framework we described in Chapter 2 is comprised of the following actions:

1. Ingesting data
2. Describing data
3. Assessing data utility
4. Designing and building refined data
5. Ad hoc reporting
6. Exploratory modeling and forecasting
7. Designing and building optimized data
8. Regular reporting
9. Building products and services

Data engineers, with their focus on data systems, generally drive the data ingestion and data description in the raw data stage. Analysts, who possess the requisite business and organizational knowledge, are often responsible for generating proprietary metadata. In some organizations, for which the data is particularly complex or messy, the generation of metadata might also be the responsibility of data scientists.

Moving to the refined data stage, data architects are often responsible for design and building the refined datasets. Data engineers might be involved if the data storage and process infrastructure requires modification and monitoring to produce the refined datasets. With the refined datasets in hand, analysts are typically responsible for ad hoc reporting, whereas data scientists focus on exploratory modeling and forecasting.

The production data stage parallels the refined data stage. Data architects and data engineers are responsible for designing and building the optimized datasets. Analysts, with the help of the data engineers, drive the reporting efforts. Data scientists, also with the help of data engineers, work to deliver the data for products and services.

Though we have been selective in our associations between job roles and actions, the reality of most organizations is that people help out wherever they can. Although data engineers typically have the deepest data systems knowledge, data architects the best data cataloging and design knowledge, analysts the most comprehensive domain understanding, and data scientists the deepest statistics and machine learning background, these skills are exclusive and many data projects can be sufficiently progressed with cursory knowledge of some of these areas.

Organizational Best Practices

In the remainder of this chapter, we discuss some best practices for coordinating the efforts across these job roles. These best practices come from a combination of our own efforts to wrangle data efficiently and from our observations of how high-functioning organizations manage their data projects.

Perhaps the most important best practice is providing wide access to your data. Of course, we are not suggesting that you provide broad access to overwrite capabilities. Rather, within of legal boundaries, everyone in your organization should have the ability to analyze the data you have. Per our discussion of driving broad data-driven value creation, your organization will benefit from opening-up access and allowing as many people as possible to find valuable insights. This initiative is more popularly referred to as *data democratization* or *self-service data analytics*. Some argue against wider access to data on the grounds that the infrastructure to support it is costly (we have seen that the generated value more than compensates for the additional costs) and that people will often find conflicting insights (which can slow down the organization while you sort them out). For this last concern, we offer two suggestions. First, build robust refined datasets and drive the majority of your analytics efforts to source from them. This will mitigate superficial conflicts. Second, embrace the remaining conflicts and build practices that address them directly. With superficial conflicts minimized, the conflicts that remain should largely represent different views on how to measure or interpret the data. It is to the benefit of your organization to uncover these differing perspectives and to find constructive ways to coordinate them. Robust insights will survive these interrogations, and your organization should trust its use in making decisions and driving operations more as a result.

Summary of Organizational Best Practices for Data Projects

The following list includes related best practices for data projects:

- Provide wide access to data
- Implement mechanisms to track data usage
- Use a common data manipulation language that spans business units and user roles
- Maintain a system that allows you to easily transition from development to production
- Consider a rotation program across roles to enable a cleaner hand-off and increase cross-functional trust

---

In conjunction with providing wider access to your data, you should implement mechanisms that can track the use of your data. This will improve your ability to resolve conflicting insights. It will also help you to determine the cost benefits of altering your refined datasets (e.g., by adding a dataset that many people are using, but sourcing from the raw stage; or by adding additional blended datasets used by many analyses). As your organization relies more and more on complex data enrichments resulting from inferences or predictions (e.g., you might regularly predict the likelihood that a customer will churn, and then use this churn prediction value to drive business operations), the ability to track the movement of your data will become critical to enhancing or protecting the utility of the inferences and predictions. In particular, the inferred values will begin to shift the operations of the organization, which will shift the data that is collected. As the data shifts, it is shifting, in part, relative to the inferences. However, if the inference-producing logic assumed that the data did not have this feedback aspect to it, the inferences might become inaccurate or biased over time. These, and related issues, are discussed in Scully et al., *Hidden Technical Debt in Machine Learning Systems* paper.[1]

Building on the idea of providing wider access to your data is finding a common data manipulation language for everyone to use. This is critical to support collaboration on analyses. At a minimum, people who want to work together on the same analysis need to share the basic tools of the analysis. Over time, people picking up old analyses to refresh or extend them will benefit from being able to immediately rerun the

---

1 Published at the Neural Information Processing Systems (NIPS) conference in 2015.

prior analysis and then to work within the same language to make their modifications. Today, many organizations rely on Excel or some flavor of SQL as common data manipulation languages and tools.

Another aspect of using a common data manipulation language is the ability to easily transition exploratory analyses to a production version or systems. Historically, many organizations have allowed exploratory analyses to be conducted in one set of tools and languages (Excel, R, Python, SAS, etc.), whereas production versions of these analyses (for regular reporting or for data-driven services and products) are often built in a more basic software engineering framework. More effective organizations have shifted toward tools and languages that support the "productionalization" (or "operationalization") of exploratory logic more directly. Most of these newer tools can simply wrap exploratory scripts in a scheduling and monitoring framework.

One final best practice: consider building a rotation program that allows people to take on the different roles associated with working with data. Superficially, it will increase the breadth of skills that your organization can take advantage of. More fundamentally, a rotation program can build empathy and trust across these job roles, leading to cleaner hand-offs of projects that span multiple groups.

Having now covered a framework for understanding data projects, how data wrangling works and fits into these projects, and how different job roles also work together in these projects, we turn our attention to data wrangling tools and languages. In line with our best practice recommendation around finding a common language, we'll cover the two most common wrangling tools and languages today: Excel and SQL. We'll also discuss a more recently developed data wrangling tool: Trifacta Wrangler. In Chapter 9, we provide a basic overview of these three tools. In subsequent chapters, we provide hands-on examples illustrating how they can be used to perform the variety of transformations and profiling involved in wrangling data.

Data Wrangling Tools

Tools for data wrangling span a number of dimensions, from general-purpose programming languages, to commodity spreadsheet applications, to visual transformation and profiling products. There are easily dozens of tools in each category, but for the purposes of this book, we're going to focus on three products that we believe represent the different focuses and strengths of each dimension of data wrangling tools: Excel, SQL, and Trifacta Wrangler. If you've worked with data in any capacity, you're probably familiar with one or more of these tools.

As we move through this chapter, we will draw distinctions between these three tools based on their supported data size, required infrastructure, supported data structures, and transformation paradigms. We want to illuminate the generic use cases for which each tool fits so that you can understand which one would best suit a particular data project.

Let's begin with a brief overview of each tool. The two most commonly used data wrangling tools are Excel and SQL. Both are more than 30 years old, which attests to the longevity of data wrangling as a task. Trifacta Wrangler, in contrast, was launched in 2012 as an outgrowth of academic research at UC Berkeley and Stanford. SQL is considered a general-purpose tool for data manipulation, and as such, is widely embedded in many relational database distributions. Excel is a spreadsheet application that allows users to manipulate, analyze, and store data in a tabular format. Trifacta Wrangler is a visual data preparation application with which users can transform structured and unstructured data and profile the output of those transformations using an interactive interface.

Data Size and Infrastructure

The first two characteristics of data wrangling tools, supported data size and required infrastructure, are very closely related. After all, you wouldn't want to use a desktop application to wrangle terabytes or petabytes of data—imagine how slowly your computer would run! On the other hand, if your total data size is only a few megabytes, investing in a big data distributed-processing platform like a Hadoop cluster would be a massively wasteful use of computing power and budget (and would likely raise concerns about your spending practices). So generally, smaller data corresponds to smaller infrastructure needs and bigger data corresponds to bigger infrastructure needs. There are exceptions to this rule, but those exceptions are beyond the scope of our discussion in this book.

Table 5-1 compares how our three tools each operate on different data sizes and infrastructures. Specifically, Excel is an application designed to run on a personal computer, whereas SQL is typically deployed on a centralized infrastructure consisting of one or more networked servers. Due to these infrastructure differences, Excel is primarily used on small- to medium-sized data (files up to one or two gigabytes), whereas you can comfortably use SQL on production transaction datasets up to the multiterabytes range (though some SQL implementations like DB2 can scale to petabytes). Trifacta Wrangler can support transforming data of various sizes—from megabytes to petabytes—by running on either a Hadoop cluster or on a single server. Trifacta Wrangler's execution environment is determined automatically at runtime based on data volume and the logical complexity of the transformations.

Table 9-1. General guidelines for the maximum data size supported by each data wrangling tool, and the infrastructure required to deploy each tool

Tool	Maximum data size	Infrastructure
Excel	MB to GB	Desktop
SQL	GB to TB	Server
Trifacta Wrangler	Unlimited[a]	Cluster[b]

[a] Trifacta provides two versions of its data wrangling product, Trifacta Wrangler. We have chosen to highlight the maximum data size supported by either version of Trifacta Wrangler.

[b] Trifacta provides two versions of its data wrangling product, Trifacta Wrangler. We have chosen to highlight the infrastructure required to support the largest possible data that can be transformed using Trifacta Wrangler.

Data Structures

Another important dimension for distinguishing between data wrangling tools is the range of structures that each data tool can handle. If you remember our discussion from Chapter 2, the structure of a dataset refers to the format and encoding of records and fields. Table 9-2 summarizes the differences between the data structures supported by each tool.

Table 9-2. General guidelines for the data structures supported by each data wrangling tool

Tool	Data structures
Excel	Grid
SQL	Tabular (uniform records)
Trifacta Wrangler	Various

Excel

Excel requires data to be laid out in a grid, though the grid does not need to be rectangular or completely filled. Often, people include multiple tables in a single Excel grid, mix descriptive text with data, or embed graphics within their spreadsheets. All of these data structures roughly conform to the constraints of the grid, but are not strictly rectangular or consistent. Within each cell of the grid, Excel supports a wide variety of value types, from numbers and percentages to dates and times. Given the level of heterogeneity that can be present in an Excel dataset, a single cell, not a record, is the most important data element in an Excel spreadsheet.

SQL

SQL expects datasets to be constructed as a set of records, in which every record contains the same set of fields. This means that any dataset that you decide to wrangle using SQL must be rectangular and must also conform to a specific schema. As with cells in Excel, the record fields in SQL can have a variety of types. Different versions of SQL support different field types, but the basic set of dates, times, strings, and numbers are universal.

Trifacta Wrangler

Trifacta, unlike Excel and SQL, can handle structured, semistructured, and unstructured data. When working in Trifacta, data does not need to be explicitly broken down into rows and columns or fully populated. Like the other two tools, Trifacta supports a variety of different data types, from the most basic integers, strings, and Booleans, to more complex custom types like dates, US states, and phone numbers.

Transformation Paradigms

The data structures that each of our tools can handle influence the ways in which users are able to transform data. We refer to this as the *transformation paradigm* for a tool.

How does the tool enable a user to define data transformation steps? Table 9-3 provides a comparison.

Table 9-3. Summary of ways each wrangling tool allows users to define transformation steps

Tool	Transformation paradigm
Excel	*User interface*: Script and wizards
	Transformation scope: Single values. User writes formulas that apply to single grid cells
SQL	*User interface*: Script only
	Transformation scope: Programmatic. User writes a script that applies to multiple records
Trifacta Wrangler	*User interface*: Script, "builder," and machine-guided transformation creation
	Transformation scope: Programmatic. User writes a script that applies to multiple records

Excel

In Excel, the basic unit of transformation is a single grid cell, which is operated on by formulas. Each cell can be referenced by using a row and column indicator. Rows are represented by sequential integers, and columns are represented by the Latin alphabet, starting at "A" and appending multiple letters as needed ("AA" follows "Z"). In this universe, "A1" is the upper-leftmost cell, "B1" is the adjacent cell to the right, and "A2" is the adjacent cell to the bottom.

The common output of an Excel formula is a single value, which is inserted into a single cell. For example, if you wanted to set the value of cell "A11" to contain the sum of cells A1 through A10, you would directly type your sum formula into cell "A11." Your sum will display in the cell, and the formula that produced that sum will be embedded in the cell itself. Of course, many transformations need to be applied to more than just one cell. Excel supports this in two ways: first, through updating references in formulas as you copy those formulas from one cell to another; and second, through array formulas. Array formulas have the added benefit of requiring the entire array of cells to be modified in a single pass, thus preserving consistency.

From a transformation script perspective, Excel supports authoring a complex series of calculations that build on one another; however, the steps in the script are embedded throughout the data. Primarily people build a series of transformation steps by creating a new column of data for each step. This practice allows you to keep intermediate steps visible. In some cases, however, people will retain just the final calculation values and delete the intermediate steps and the formulas used to generate the final values. In this case, unless you create external documentation, the dataset itself does not preserve any lineage or history showing how values were created.

SQL

SQL operates at the level of entire records. Although it is possible to isolate a specific field within a specific record, more common uses of SQL involve appending or overwriting records to an existing dataset, or selecting subsets of records and record fields for calculations like aggregations. SQL enforces encoding restrictions on record fields; this means that all of the values in a single field must have the same datatype. A column with a datatype "integer" cannot contain fields of string values. Most SQL systems also require the designation of primary keys for each record, and enforce the unique occurrence of these keys (i.e., the dataset will contain at most one record with any specific primary key value). Any complete transformation statement in SQL requires specifying the record fields involved in the transformations, their output locations (by name or position within the record), as well as every other field that is "passing through" (i.e., fields that you want to keep but that do not require transformation). In other words, every complete transformation in SQL requires specifying the entire schema of the dataset.

In SQL, you can write a transformation script as a single long (and often nested) query or as a series of queries. As already noted, a central difficulty with SQL steps that build on one another is that they require full specification of the fields that you want to carry through the calculation. This results in verbose, and in many cases redundant, information in a SQL transformation script. If the script is written as a series of queries, storing intermediate results along the way, the user can interrogate the intermediate values to assess the validity of the transformations (though managing these intermediate datasets becomes an additional task, in this case). If the script is written as a single, long, and nested query, it is difficult for users to create and profile intermediate results. Doing so requires creating valid variants of the query that output the intermediary result versus the final.

Trifacta Wrangler

Trifacta Wrangler also operates at the level of record fields and at the level of entire records. In many cases, data transformations specified in Trifacta Wrangler look like subsets of SQL query statements. The common subset will specify just the record fields involved in the transformation as well as how they are combined—fields that pass through do not need to be specified (whereas in SQL they do). These abbreviations result in more readable transformation scripts (relative to SQL). Whole records can be operated on from the perspective of filtering out records or of aligning the fields across records from different datasets during enriching transformations like joins and unions.

Trifacta transformation scripts place all the transformation logic in one place, so users do not need to click around on cells or blocks of cells to see what logic was applied. Trifacta transformation scripts also support abbreviation of passed-through record fields, streamlining their readability, as well as the ability to click back through script steps to see the state of the dataset at intermediate points.

Of course, profiling is a critical task coupled with specifying data transformations. Excel supports reviewing of individual cells as well as automated validation of these cells. Simple distributional checks are supported for numeric blocks of data (e.g., sums, averages, minimums, and maximums) but not for strings or more complex values like lists and arrays. Visual profiles of the data (e.g., bar charts and scatterplots) can be created but are time consuming. In SQL, some modern tools like Periscope and Looker connect data visualizations directly to SQL queries. However, most SQL users profile the data by instantiating the dataset (either in intermediary form or as final result) and then applying simple SQL queries to view and summarize the data.

Trifacta embeds profiling into the core user interface of the product. Two primary visual charts are available in the product: one illustrates the validity satisfaction of individual record field values, the other illustrates the distribution of values to give the user a sense of the "shape" of the data. Furthermore, the charts are interactive, allowing the user to quickly select subsets of values to create transformations targeted to that subset.

Choosing a Data Wrangling Tool

So, which data wrangling tool is best? Although we obviously have personal preferences, there is really no single correct tool choice. Your choice of data wrangling tool depends on what you hope to do with your data. We emphasized data volume, infrastructure scale, and data structure as key differentiators for data wrangling tools, as well as elements of your data project that you should consider before selecting a tool.

About the Authors

Tye Rattenbury is a technical advisor to Trifacta and was previously Trifacta's lead data scientist. He holds a PhD in Computer Science from UC Berkeley. Prior to Trifacta, Tye was a Data Scientist at Facebook and the Director of Data Science Strategy at R/GA. Tye currently leads a Data Science and Machine Learning team at Salesforce.

Joe M. Hellerstein is Trifacta's Chief Strategy Officer and a Professor of Computer Science at Berkeley. His career in research and industry has focused on data-centric systems and the way they drive computing. In 2010, *Fortune* magazine included him in its list of the 50 smartest people in technology, and *MIT Technology Review* magazine included his Bloom language for cloud computing on its TR10 list of the 10 technologies "most likely to change our world."

Jeffrey Heer is Trifacta's Chief Experience Officer and a Professor of Computer Science at the University of Washington, where he directs the Interactive Data Lab. Jeff's passion is the design of novel user interfaces for exploring, managing, and communicating data. The data visualization tools developed by his lab (D3.js, Protovis, Prefuse) are used by thousands of data enthusiasts around the world. In 2009, Jeff was included in *MIT Technology Review*'s list of "Top Innovators under 35."

Sean Kandel is Trifacta's Chief Technical Officer. He completed his PhD at Stanford University, where his research focused on user interfaces for database systems. At Stanford, Sean led development of new tools for data transformation and discovery, such as Data Wrangler. He previously worked as a data analyst at Citadel Investment Group.

Connor Carreras is Trifacta's Manager for Customer Success, Americas, where she helps customers use cutting-edge data wrangling techniques in support of their big data initiatives. Connor brings her prior experience in the data integration space to help customers understand how to adopt self-service data preparation as part of an analytics process. She holds a BA from Princeton University.

Colophon

The animal on the cover of *Principles of Data Wrangling* is an angular crab (*Goneplax rhomboides*), so named for the sharp corners of its carapace. It lives in the Atlantic Ocean along the coasts of Europe and Africa, as well as the Mediterranean Sea. These crabs dig complex burrows in muddy sand, and also are found in shallow water to depths of about 325 feet. They are colloquially known as "mud-runners" due to their evasive maneuvers when found on the beach.

The species' color varies, including shades of yellow, orange, pink, and red. On average, their trapezoidal shells are 1.5 inches wide. The chelipeds (the limbs with a pincer at the end) are much longer on males than on females: sometimes up to five times the length of the carapace. Angular crabs have retractable eyestalks.

Crabs are omnivores, feeding primarily on algae, but also mollusks, worms, fungi, and other crustaceans (depending on what is available). They moult many times as they grow to adulthood, taking in water to expand and crack open their old shell. Once this is achieved, they spend several difficult hours getting free, and then must hide until the new shell hardens.

Many of the animals on O'Reilly covers are endangered; all of them are important to the world. To learn more about how you can help, go to *animals.oreilly.com*.

The cover image is from *Wood's Illustrated Natural History*. The cover fonts are URW Typewriter and Guardian Sans. The text font is Adobe Minion Pro; the heading font is Adobe Myriad Condensed; and the code font is Dalton Maag's Ubuntu Mono.

Learn from experts.
Find the answers you need.

Sign up for a **10-day free trial** to get **unlimited access** to all of the content on Safari, including Learning Paths, interactive tutorials, and curated playlists that draw from thousands of ebooks and training videos on a wide range of topics, including data, design, DevOps, management, business—and much more.

Start your free trial at:

oreilly.com/safari

(No credit card required.)

Milton Keynes UK
Ingram Content Group UK Ltd.
UKHW030242281123
433366UK00011B/225